Exploring
Microsoft®
Office
XP

Maximize Your Productivity with the Newest Features in Microsoft Office XP

John Breeden II
Michael Cheek

Exploring
Microsoft®
Office
XP

Maximize Your Productivity with the Newest Features in Microsoft Office XP

John Breeden II
Michael Cheek

PROMPT®
PUBLICATIONS

PROMPT© Publications is an imprint of Sams Technical Publishing, 5436 W. 78th St., Indianapolis, IN 46268.

International Standard Book Number: 0-7906-1233-X
Library of Congress Catalog Card Number: 2001089991

Acquisitions Editor: Alice J. Tripp
Editor: Kim Heusel
Assistant Editor: Cricket Franklin
Typesetting: Kim Heusel
Indexing: Kim Heusel
Proofreader: Patrick Brady
Cover Design: Christy Pierce
Graphics Conversion: Christy Pierce
Illustrations: Courtesy the authors

PRINTED IN THE UNITED STATES OF AMERICA

9 8 7 6 5 4 3 2 1

Contents

About the Authors

The Senior Editor and Lab Director for *Government Computer News,* John Breeden II is in charge of advanced computer product testing. His reviews appear weekly in *GCN* and in the Thursday Business Section of *The Washington* Post. Prior to joining *GCN*, he wrote a computer column called "On The Chip Side" that had a circulation of 1.1 million in newspapers throughout Maryland, Virginia,

John Breeden II

Pennsylvania, and West Virginia. He is the vice president of Noble Order Press Enterprises and editor-in-chief of Gameindustry.com. He is the co-author of *Guide to Webcams*, a book that explains how to set up and maintain a Web camera for both business and personal use.

A professional journalist for 14 years, Michael Cheek began his adventure with computers at the age of 15, when the personal computer revolution was in its infancy. Through the years, he has written about thousands of computer products. Cheek's reviews have appeared in *The Washington Post,* the *Atlanta Journal-Constitution*, *Government Computer News*, and on GameIndustry.com.

Michael Cheek

Cheek currently works as managing editor for iDEFENSE, a cyber-security firm that provides rapid intelligence to information technology professionals worldwide. Prior to joining iDEFENSE, Cheek was the lab director at *GCN*.

Dedication

Gracious thanks to Karen, who let me work away the hours looking at Smart Tags when we could have been having fun.

John Breeden II

Thanks to my supportive network of friends, family and colleagues, who can search through these pages and find subtle salutations (and sorry if I forgot you). A special thanks to Todd A. Brown, CPA, who's more than just my accountant.

Michael Cheek

1

Countdown
to Launch

*Office XP may go
down as one of
Microsoft's best
improvements
ever.*

Resistance is futile. Just when you thought you had finally mastered Office 2000 or Office 97, Microsoft unleashes its newest version of the most dominating office suite in the universe. Your first urge may be to resist, but should you?

An upgrade to Microsoft Office XP will come eventually. We suggest embracing it with gusto! Office XP may go down as one of Microsoft's best improvements ever. It is sure to change the face of workplace productivity.

Be the first on the planet — or at least your block — to tap into the new features Office XP offers. Heck, wouldn't you like to figure out some of the features in the older versions?

This book will help you become a master of this Office domain in no time.

This guide does that.

Office XP stands for Office eXPerience. This could be taken to mean that when using the suite, you are having the Office eXPerience, or that with all the new features, you will need some eXPerience to make use of all the features of the applications. For our purposes, let's call this guide the Office eXPloration eXPerience, as we plan to guide you though the entire suite of applications, every step of the way, with heavy emphasis on new features that nobody has eXPerienced yet because many are totally new. And we promise that is the last time we will use the word eXPerience in this book with the funny Microsoft capitalization.

While the XP moniker is used for the suite, you'll find many of the applications retain the year appendage. Those of you using Office 95, 97 and 2000 are familiar with the year. This version is known in some circles as version 2002. When burrowing into the depths of the applications, you might run across something that says Word 2002 or Excel 2002.

But don't let the marketing and version lingo bother you. By reading this book and following its easy-to-learn techniques, you will increase your proficiency with Microsoft Office by whatever name you call it. Our goal with this book is to give you the knowledge necessary to unlock the many features of this complex, yet highly useful suite of applications.

When first loaded, Office XP may appear like something from a parallel universe — somewhat familiar, but just different enough to confuse you. This book will help you become a master of this Office domain in no time.

To their credit, engineers at Microsoft have done an admirable job of revealing some of the features of Office XP. Navigation paths within the suite have been greatly cleaned up compared to previous versions, and applications now alert you almost every time an option is available.

Microsoft has also added a lot of new features to the mix. The emphasis on communications in the business world — especially over the Internet or an Intranet — could not be ignored in any office suite and certainly not in the suite used by about 90 percent of the general public.

These new features are really premiering in Office XP, as most other office suites have not approached this level of complexity.

As the authors of this book, we have watched office suites develop for at least 15 years. Fifteen years! That's a millennium in the computer world, but we've written more than one term paper on an Apple II computer using AppleWorks, and on archaic PC applications.

This book assumes the reader has at least a passing or casual knowledge of office suites in general and, hopefully, some familiarity with a previous version of Office, be it Office 95, 97, or 2000. We won't go into too much detail about how to create a document, for example. We assume most folks can go to the **File** menu and select "**New...**".

Instead, we want to concentrate on the latest features so readers can get the most out of their investment in the applications, both in terms of time spent and quality output created.

We will dwell on some of the totally new features of the applications such as Office's new Web site. Since this

This book assumes the reader has at least a passing or casual knowledge of office suites in general.

Of all the applications a computer can currently run, an office suite is probably the most useful.

is a brand-new component, no one has had experience with it. But occasionally, we'll provide some insight into existing features and even some of the basics, such as how to add page numbers in Word, how to change the colors on an Excel chart, and much more.

Short of doing the programming ourselves, we have been involved with the applications throughout the development process, right from the first quirky beta Microsoft released up to the gold master that was eventually duplicated into the final product.

The Microsoft public relations team deserves special thanks for keeping us informed every step of the way and for passing our feedback on the applications directly to the Microsoft Office engineers doing the design work. We'd like to thank Keith Hodson, our local man from Microsoft. A very special thanks goes out to Jennifer Flentge and Anne Eddy, our primary PR contacts on this project, and David Jaffe, who was our inside man at the company and someone whose knowledge of Office version plans proved invaluable. All these folks are exemplary in their duties and would be a credit to any organization.

Time Travel: A Brief History of the Office Suite

It's truly an amazing world.

Technology has advanced to a point where the seemingly mundane tasks we perform every day would bewilder our grandparents and would be considered nothing short of mythical to ancient peoples. But for us, it's just another day at the office.

Of all the applications a computer can currently run, an office suite is probably the most useful. At your fingertips, Office provides more power than Gutenberg could have ever imagined. More than just converting spoken words to paper, a suite can actually process those words, a concept that has only recently come to be used and understood in the mainstream. The question of "how can you process words?" surely would have greeted anyone trying to use the term "word processor" even 20 short years ago. Back then, peaches and soup were processed, but not words.

This is not to say words were not important, but computers have opened up the world of tailored words, where different tools and tricks can add emphasis to meanings.

Words can be processed in that they can be shaped, colored, and arranged in such a way as to maximize their effectiveness. The key to this is ease of use. If you are typing a document on a typewriter and you suddenly think your second paragraph would be more effective as your opening line, you have little choice but to start again. On a word processor, however, a simple cut-and-paste does the trick.

Office suites can also do a myriad of other useful tasks such as keeping track of your schedule, balancing your books, tracking all your contacts via physical "snail mail" and e-mail addresses, creating stunning presentations that go beyond even what a processor of words can accomplish, building dynamic and tempting Web pages, enhancing and changing photographs, and even directing you to a good Chinese restaurant.

Office suites can also do a myriad of other useful tasks.

In the earliest computer days, office suites did not exist.

Whew! All that and they cost less than a personal assistant. But to truly appreciate how powerful a jumble of applications Office XP really is, it's good to know the history of office suites in general. Suites weren't always so pretty or functional and only relatively recently have been created out of a bundle of disparate tools.

In the earliest computer days, office suites did not exist. Functional programs that did one task or another could be bought separately. WordPerfect made WordPerfect. Lotus released 1-2-3. Microsoft provided MS-DOS. Each company had its own specialty.

On one hand, users only had to purchase whatever component they needed. An accountant might purchase a spreadsheet application like 1-2-3. If you were creating databases, you would buy dBase. And any wordsmith knew WordPerfect.

But back then, each application alone cost almost as much as the suites of today. Moreover, just because you do most of your work with a spreadsheet does not mean that you will never need a word processor. In order to write a letter, you would be required to buy a word processor, even if you only needed to use it once. Finally, the applications lacked compatibility. Getting your information from one to another was nearly impossible, even among applications of a similar nature.

In some cases, data could be shared in limited fashion by saving data as text. But any formatting in the document was either lost or appeared as unintelligible gibberish.

To a very small degree, the "buy only what you want" mentality survives. Even in Office XP, several applications are available for purchase separately. These applications

— primarily PhotoDraw, Publisher, Visio, MapPoint, and Project — can integrate into Office XP, but don't come with any of the editions. We cover these applications in Chapter 9 just in case you decide you want to use them. Mostly these are specific applications that most people would not use, but are very important to the folks who do need them.

At the same time as these individual advanced applications evolved, other smaller applications were being bundled together. Generally thought of as being crippled without the features of the more expensive full-blown applications, these were the earliest suites. Microsoft Works remains among the granddaddies and is still available today in an updated format.

More than 15 years ago when the PC revolution began, AppleWorks appeared. Apple II computer users everywhere used 5.25-inch floppy disks — that were truly floppy back then — and loaded this little suite.

But sometime after the "works" suites, around 1993, someone put together all of the full-blown applications into their own suite. Back then, the applications were disjointed and didn't exactly fit together. And they came on dozens of floppy disks.

When the personal computer started to make inroads into homes and businesses, the obvious comparison was to a typewriter.

Words on paper

The oldest component of the suite would probably be the word processor. When the personal computer started to make inroads into homes and businesses, the obvious comparison was to a typewriter. After all, modern keyboards are pretty similar to the old typewriters. Plus, when a person types, the words appear on the monitor, just like they used to do on the paper.

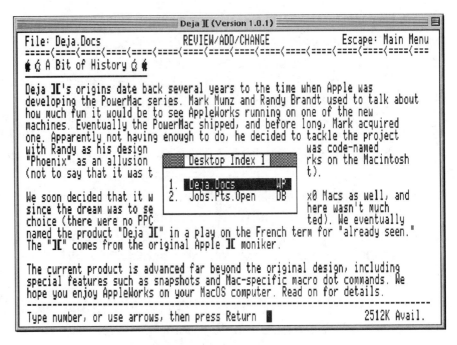

Figure 1.1

The earliest of word processors actually started to appear on advanced typewriters.

The original word processor was actually not a computer at all, or at least not a personal computer. The earliest of word processors actually started to appear on advanced typewriters. The devices were primitive by today's standards, but at the time, they offered an advantage that few had ever experienced before: the ability to correct words or even entire sentences after they had been typed.

These devices had what became known as "lookback" memory or a cache. With a very low-end processor, the typewriters could capture a few characters, words, or even sentences and display them in a tiny window. If a mistake was noticed, it could be corrected before it got onto the relatively unforgiving paper. Folks started thinking, "If I can see a few words back, why not an entire paragraph, or even my entire document?"

Dedicated word processors emerged as somewhat huge boxes that combined a keyboard with a mostly green or amber cathode-ray tube display. They were as heavy as a stack of bricks, but they allowed users to proofread entire documents before printing them out. They could even save documents — originally to memory and later to disks — for editing later.

Then came the PC revolution.

Computers had to do more than word processors, though, because they cost five or six times as much. Back in those days, buying a computer was a lot like buying a car. Potential buyers would actually visit computer showrooms featuring all the latest models. Buying a computer was an investment, and users wanted to get the most out of it.

So, the word processor moved to the PC. In many ways people really liked the early word processors. They were far from the complicated, instruction-heavy, generously proportioned feature monsters we have today.

The PC's first word processors were all character based, much like the behemoth dedicated word processors. WordPerfect personified it with a blank blue screen and white text.

But as word processors advanced, the feature known as WYSIWYG emerged. This cute phrase, pronounced wiz-see-wig, meant that whatever you saw on your screen would translate exactly to the printer — *what you see is what you get*.

Buying a computer was an investment, and users wanted to get the most out of it.

It took time for programmers to tap into the power of the computer.

The problem here was twofold. First, WYSIWYG, often wasn't. Almost all printers back then were dot matrix that lacked accuracy. Page breaks on the word processor often did not translate into actual page breaks, which led to a lot of printed lines directly on top of the area where the perforated pages were to be ripped.

The other problem with WYSIWYG word processors is lack of control over the text. Other than basic formatting options such as setting the position of the text on the page, you could not do a lot. It took time for programmers to tap into the power of the computer.

When they did, processors began to look more like the sleek applications of today. In this world, the complicated codes of the word processors were second nature. For example, if a sentence was supposed to be bold, users had to place their cursor at the beginning of the sentence and type a code like [/B+] and then type [/B-] at the end. The computer would read the left bracket and know that a code was about to begin. Then slash and B would alert it that it is supposed to do something with boldface text. The plus and minus would tell it to either begin or end the bolding. And finally the right bracket would tell it to stop reading the code and go back to printing text. Second generations of these word processors would actually display the text on the screen how it should be formatted after the code, bringing some elements of the WYSIWYG interface to the table.

A lesser-known fact is that the word processors of today are still littered with those same codes. Only the interface has changed. When you press the **B** button in Office XP for example, the text will begin to bold itself. But, a code will be inserted at the beginning of the bold text, just like in the old days. The only difference is that the codes are hidden. An explanation on how to find and use these codes to your advantage is found in Chapter 3 about Word.

Crunching numbers

A close second in age and popularity in any suite is the spreadsheet. There have never been any dedicated spreadsheet computers — at least that we have ever seen — but try to run a business without a spreadsheet today.

Unlike word processors, spreadsheets changed little as they advanced from generation to generation. Other than simplifying the interface somewhat, they remain largely unchanged. Now, spreadsheets simplify complex mathematical calculations and make table building and chart creation easy.

What has changed is how spreadsheets are used. They're not just for accountants, statisticians, or the mathematically inclined. Spreadsheets organize data.

Excel, the spreadsheet component of Office XP, comes from a long history of top-notch spreadsheets. In head-to-head comparisons over the years, Excel has almost always come to the top of the heap based both on computational power, speed, and the ease of the user interface.

The current version of Excel is advanced in two important ways. First, it has a larger capacity than the early spreadsheets even doubled or tripled. You can literally put the entire U.S. government budget — millions of lines — into an Excel spreadsheet. If your personal or busi-

There have never been any dedicated spreadsheet computers, but try to run a business without a spreadsheet today.

ness finances are larger than that, then please contact us because we would love to work for you.

Secondly, it adds a ton of new features that make data analysis truly comprehendible, even on a very large scale. Basic and even advanced formulas are easy to enter and pivot tables make graphical representation of data a breeze.

Electronic mail now follows light-speed serpentine paths to reach a destination.

Office XP also adds a new definition to spreadsheets, that of an artificial intelligence. When you paste data into the spreadsheet, it tries to figure out what type of data it is, and attempts to display it properly. Office XP now allows "living documents" that change as the information they are based on does. Never before has a spreadsheet held so much potential. We examine this in much detail in Chapter 4 on Excel.

The spreadsheets can also be tightly integrated into the Office XP database application, Access. Access is a powerful database that is fully explored in Chapter 7.

At dot-com this or that

E-mail has only recently emerged in the last five years as part of an office suite thanks to the whole Internet revolution.

Electronic mail now follows light-speed serpentine paths to reach a destination. The occasional delay in communication equals seconds or at the very worst minutes, not hours and days.

Outlook, the e-mail application in the suite, actually uses Word. If you master Word, you have all but mas-

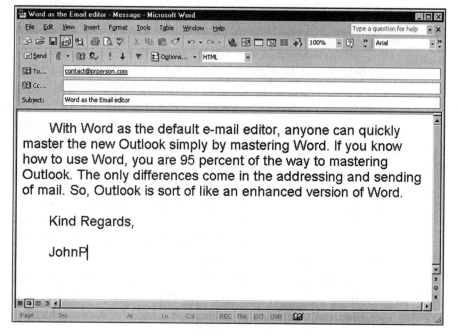

Figure 1.2

tered writing an Outlook e-mail *(Figure 1.2)*. We cover exactly how to generate professional e-mails using the new Office XP tools in Chapter 5 on Outlook.

We also cover scheduling and contact information, now an essential part of any personal information manager.

Presenting the universe — in style

The newest tool in the package of all office suites is presentation software. Business turns on a good presentation today, which is nothing new. But the computer has become an essential part of this process. Projectors have made large-screen presentations possible in almost any

Projectors have made large-screen presentations possible in almost any environment.

PowerPoint has always been the most popular application in this realm.

environment. The projectors, which display any information on a computer on the wall or a screen, are now mostly portable, so folks are starting to carry them when they are planning a presentation.

Even without a large projection device, the standard screen sizes on laptops are getting larger. This means that a few people sitting reasonably close together can watch a presentation conducted on the laptop. It is quick and easy to make a basic presentation, save it to the hard disk of a notebook, and take that computer on the road.

As in the realm of spreadsheets, Microsoft has always had the upper hand when it comes to presentation software. PowerPoint, the presentation software bundled with Office XP, has always been the most popular application in this realm.

Of all the changes in Office XP from previous versions, it is fairly safe to say that PowerPoint has had the biggest facelift. Both the static elements of the application such as charts and tables, and the active elements such as movies and sounds, have been revamped to the point that they are really, really different from previous versions.

We think veteran presenters will find that they can still produce standard presentations with PowerPoint, but that the multitude of new features will make presentations sing. This is truly a breakout product and knowledge of how to find and use all the new elements in PowerPoint will be very helpful in both creation and presentation.

Chapter 6, the largest in the book, is devoted entirely to PowerPoint and how to make and use presentations. We think this will be helpful to beginners since about half the new features were designed for ease of use. This should help experts, too, since the other half are designed to improve presentations and make them easier to create.

So, presentation tools are about the latest addition to the office package, but they are quickly becoming one of the most important.

Spinning the Web

The final component in office suites is also relatively new. The Web page design application, at one time an afterthought, is now solidly entrenched in most office suites.

FrontPage — the Web page creation tool — may have gotten off to a late start in this arena, but it's become one of the most popular gaining mainstream acceptance, in part because it makes Web design a fairly simple matter. Some Webmasters don't even learn the raw HTML (hypertext markup language) that the early Webmasters relied on. They don't have to because FrontPage does all the coding for them. Microsoft has been accused of designing FrontPage so that it does not work optimally with non-Microsoft browsers, such as Netscape Navigator. These claims have not hurt FrontPage's popularity, however, and Microsoft smartly added it to the suite.

In Chapter 8, we will go over the pros and cons of using FrontPage and zero in on some of the improvements over previous versions.

The Web page design application is now solidly entrenched in most office suites.

What is Office? Which Office?

The Microsoft Office suite line has always been one of the most popular in the world. Partly because of the Microsoft name and partly because of shrewd marketing tactics, Office has been top dog in this area. There have been several challenges to the crown as of late, most notably by Lotus SmartSuite Millennium, a package owned by another giant: IBM Corp.

Office has always boasted tight integration among its many components

But Office has always boasted tight integration among its many components — a factor that has helped it to stay on top. Users like the fact that Office maintains the same look and feel throughout the application and, for the most part, that data is 100 percent compatible among applications.

That said, many of the Office upgrades in the past have been mostly superficial. It was more like adding a fresh coat of paint to an old house instead of adding on a few more rooms. Since all versions of Office are downward compatible — and continue to be — users did not feel the need to purchase each and every version of the suite as it was released.

Some did, but for the most part, if a user had a stable version of Office 97, he or she would rarely upgrade to Office 2000. Especially among the professional versions of the software, the price tag is quite high, which led many consumers to stick with the older versions that had been proven on the workplace battlefield. In this manner, Microsoft was in competition with itself, as it had to convince users that they should spend new money on an upgrade.

Office XP is the first real version of the suite that really does break the mold. It adds new rooms to that old house, puts a lot of new furniture into existing rooms and puts a new splash of paint on there as well.

What Microsoft did was to gather together focus groups of users and ask them what they liked about the suite's applications, what they didn't like, and what they would like to see added. What they discovered was what users had been telling the authors of this guide all along — that Microsoft Office was the preferred office suite, but that it had a lot of holes in some areas that needed fixing.

In general, the problem areas boiled down into two old complaints and one new request.

> *Office XP is the first real version of the suite that really does break the mold.*

So what's new in Office XP?

In the area of old complaints, users said they knew that Office had a lot of nice features, but they were difficult to find. When we have written miniguides and previews of the Office suites of the past, much of the time was spent telling users how to drill down three and four menus to find helpful features.

Truthfully, if Microsoft had done a perfect job of making features in Office XP known, there would be less of a need for this book. However, Office XP does do a good job of highlighting and alerting users when choices can be made.

SMART TAGS

In Word, a hard return will automatically result in a tab being placed at the beginning of the next line. However, if a user moves the cursor around, as in the case of editing the document, and inserts a return, a little lightning bolt appears. Users who click on the bolt will be asked to define the new tab rules. You can tell the program to ignore the seemingly new format request, chalking the abnormal typing up to human error or whim.

The little lightning bolts are called Smart Tags *and are integrated throughout the Office suite.*

The little lightning bolts are called *Smart Tags (Figure 1.3)* and are integrated throughout the Office suite. It's a way of forcing users to learn the program and what is possible when using Microsoft Office XP. This sort of stealth tutorial is great for new users, as long as the constant prompting does not become annoying. In the end, it should lead to users who are more informed about how to format and change their documents so what is reproduced on the screen and eventually on paper comes pretty close to what was originally envisioned.

Figure 1.3
Smart Tags

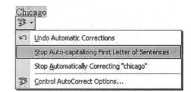

The other area of primary concern expressed by individual users is the lack of control over the things that happen automatically within the suite. A good example is when a user types a URL like http://www.gameindustry.com and Word automatically by default makes the address a hyperlink. If a user is creating a document for the Web, this is great, since readers can click on it and be taken directly to the site in question. However, this looks kind of silly when created for a document that is supposed to be put on paper. In the past, most users really had no idea how to control these automatic occurrences, as the controls were often buried several menus down and in different places.

Office XP improves on this process by triggering that same lightning bolt Smart Tag whenever an automatic process happens. From the drop-down menu, users can remove the automatic process in that single incident or in all future occurrences. So you could, in the example above, select "stop making hyperlinks" and not run into that problem again.

Another area of concern expressed by all groups was data recovery.

CRASH GUARD AND DATA RECOVERY

Another area of concern expressed by all groups was data recovery. Unless you work in a very secure area with power strips, uninterruptible power supplies, and computers that can't crash — yea, right — then data loss is an important topic. In the past, suites have done an admirable job at data recovery, but even if the majority of a document following a crash could be saved, oftentimes the most recently added data was lost.

Microsoft has fixed this for the most part by really punching up the data recovery part of all the components within the suite. The first thing it has done is to implement a crash guard. When an application becomes unstable, Office attempts to save what it can.

But during events like sudden power failures, no software program in Microsoft Office can help save data. For this, Microsoft has implemented some good data-

The recovery software that comes with Office XP is very accurate.

recovery technology. When users get back into the system following a crash, they are given a list of files that have been saved or can be recovered. There could be multiple examples of the same file in some cases.

For example, when logged back in, users will see the document that was being worked on prior to the crash. The last saved version of the Microsoft Office program will be listed, which is the last time the user actually went into the File menu and forced a save or the last time the software did an automatic save. Also listed will be the recovered version of the document, which may have additional information that was added after the last save.

The recovery software that comes with Office XP is very accurate. In tests where we forced a crash by pulling the plug on a system, about 96 percent of the time all the data that was added, even following the last save, was recovered. On occasion a few words or even an entire sentence might be missing, but for the most part the protection is very good.

Fortunately, most users will not have the opportunity to experience the brilliant crash-recovery software, but it is nice to know that it is there, kind of like a lifeboat on a cruise ship. If you have to use it, you are in a bad situation, but you would be a lot worse off without one. Businesses and individuals who are working on mission-critical applications will find this feature very helpful.

A more hidden advantage to this detailed crash guarding process is that most users should be able to recover from a crash on their own, which will put less of a burden on the information technology support staff.

As part of the error-checking process, users have the option of sending error reports back to Microsoft. The logic behind this is that if several people in Excel, for example, experience crashes when trying to add a color to a chart, Microsoft can go back and fix the problem. In the past, Microsoft engineers have complained that they spent a lot of time fixing problems from single users who screamed the loudest, when in fact large groups of quiet sufferers with a different problem were ignored.

These updates are distributed via the Internet. For users afraid of Big Brother watching, they can click on another button to see the contents of the report being sent to Microsoft. There is no personal information inside this report, which is mostly just the status of what Microsoft Office program was running and what function caused the crash. Users also have the options of not reporting the error, though we would strongly recommend sending the report so a patch can be generated.

In the past, speech-recognition programs have been less than accurate.

SPEECH RECOGNITION

Office XP also integrates an accurate version of speech technology into the program. This lets users move within the suite without touching the keyboard and enter data into some of the programs. In the past, speech-recognition programs have been less than accurate, with the best achieving perhaps 90 percent correct recognition after extensive training.

The Microsoft speech component to Office XP is good right out of the box. Especially for small, easily understood tasks, no training is really required. And when you do have to train for more precise recognition, the training times are quick. This will be a boon to disabled users and those users who are very slow when typing. Even the fastest typing professional might see some improvements if he or she doesn't have to reach for the mouse and start clicking through menus to format or navigate within a document.

The Microsoft speech component to Office XP is good right out of the box.

Remember that a good soundcard and a microphone are required. No matter how good any speech recognition is, it won't perform well if the input device — in this case the microphone — is cheap and can't pick up subtle sounds. We'll go into more details in Chapter 10 on speech.

TASK PANES

The other major improvement for most users is going to be the use of Task Panes. Task Panes are frequently used commands that can be placed in their own windows. These little windows can be moved around the screen and dropped into out-of-the-way, yet accessible areas on the main screen.

A good example of using a Task Pane to save time would be when formatting a document in PowerPoint. Once the data for a new slide is complete, to format it users can just click on the example that was used in the last few slides — stored in the pane — and the program will automatically lay out the new page the same way. This can save multiple steps when creating a presentation, especially one where the slides are basically homogeneous. Formatting properties in other programs, search windows, and even the new file menu can be made into a Task Pane. Users with large monitors will benefit most from this improvement, as the panes can be set off to the side of the work area.

COLLABORATION TOOLS

This version of Office truly embraces the Web. Using the Internet or a company intranet as the backbone, users can be truly collaborative within Office XP. Schedules can be shared and documents can be posted for sharing and dissemination.

Along the lines of collaboration, a chief complaint from groups of people who were already collaborating was that the main tool of sharing data, giving feedback to one another, was too difficult to use. Both authors have seen this to be true firsthand.

In previous versions of Office, users could send documents via e-mail to another person for editing. When the person who was doing the editing would make a correction, the changes would be displayed in a different color. If they tried to delete something, it would just strikeout the old letters, leaving the original letters intact. So when the original user got the document back after feedback and editing were conducted, it was a rather tangled mess of multiple-color text and strikeout characters.

This is one of the biggest improvements in the suite that is not an all-out new feature. The feedback component is greatly enhanced. Instead of all those strikeout characters and funky colors, corrections are now made and placed in the margin of the document. In this manner, it is a lot closer to an actual paper edit, where an editor will circle what is wrong and then write in the changes.

When the original user gets the document back, all he or she needs to do is travel down the margin of the document and find all the corrections. Office will do this

Schedules can be shared and documents can be posted for sharing and dissemination.

automatically for the user, but it is just as simple to do it by hand, since it's just a matter of scrolling and following lines back to the correction point.

Microsoft also embraces a Web community page.

Hosted by a number of Internet Service Providers or put right on the company intranet, the internal shared workgroup page is a way that workers in a department can communicate in a relatively open format.

How it works is that users within the workgroup are given access to a Web page, on the Internet or on an intranet, through the Office XP software. The site itself can be open — meaning that anyone with the Office software can access it — or it can be password protected so that only authorized users can get in. It's fairly secure, but due to the nature of the collaboration, it is best if company confidential information is kept off the page.

What the page seems designed for is meeting planning and feedback among members of the group that have access. A running schedule can be placed there, for example, and group members can check the site each day to see if meetings have been scheduled or if existing meeting times have changed.

The page can be set up any way users want, but there is a place where a chat board can be placed. Users can post problems or ideas that can be read by anyone else with access to the site. Those readers can then respond to the post and it moves from there. This is probably a better discussion forum than e-mail, because it's likely that not everyone in a large workgroup is going to care about every question or incident. So instead of forcing users to read e-mail sent out to the entire group, they

Users can post problems or ideas that can be read by anyone else with access to the site.

could log onto the company or group page and just read the posts that they think pertains to them.

Since the board can be put on the Internet somewhere, users in groups dispersed throughout the country can still have a place to meet regardless of physical location. This part of Office is a nod to the rapidly expanding field of communications.

Blast Off

Office XP has come a long way in a short period of time. Most industry professionals did not expect to see so much from Office XP, considering the somewhat cosmetic improvements seen in previous versions of the software. Microsoft has thrown everything but the kitchen sink into Office XP, and somehow managed to integrate everything together so that it is both compatible and homogeneous.

If learning Office XP sounds like a daunting task, well, we won't kid you; it's not going to be easy. But it should not be that difficult, either. You might consider first reading chapters of this book pertaining to applications you are going to be using, though it is a good idea to familiarize yourself with all aspects of the suite. Sometimes certain applications might work better than the standard ones in terms of the task you have at hand. Also, after reading each chapter, it might be a good idea to practice what you have learned. It's the best way to commit all of this newfound knowledge to memory.

So, onward and upward to Office XP.

Microsoft has thrown everything but the kitchen sink into Office XP.

2 ◆ Get Ready for Launch!

Installing Office XP

System Requirements

Installing Office XP appears to be relatively simple. But it takes a little more than a few clicks to install Office XP correctly. Otherwise, it can be like a jet pack without enough fuel — you could be working in Office at full speed when suddenly you're brought to a dead stop because you need a component that wasn't installed.

Before you attempt to run Office XP, however, make certain your PC has enough power. Otherwise, you won't have the boost needed to launch Office into orbit.

A witticism in the computer industry goes something like this: the marketing staff writes the system requirements, while engineers write the applications.

What that means is that the marketing people prefer that an application work on every computer, thus making the product a "must have" for everyone. This increases

Before you attempt to run Office XP, make certain your PC has enough power.

27

sales and profit. The engineers want to add the coolest and best features regardless of system requirements. The clash comes if marketing stretches the system requirements to expand the market a bit. We've found that the PC with the "minimum requirements" won't result in a crash, but it often will result in lackluster performance, bordering on unsuitability — and a very unhappy user.

Office XP runs fine under any of the operating systems supported by Microsoft.

In recent years, a consumer-driven "truth in systems specs" campaign resulted in more realistic requirements on the packaging. But it's safer to assume that you will need double the minimum requirements listed on the package to get the most out of the optional configuration. Microsoft is no exception to this rule.

In order to determine exactly what the system specifications for Office XP are, we installed the program on several computers, some with the recommended minimum requirements and some with slightly higher or even lower specifications.

Office XP runs fine under any of the operating systems supported by Microsoft. It works under Windows 95, 98, and Windows Me. It also ran well under Windows NT 4.0 Workstation with Service Pack 5 or above and Windows 2000 Professional. Basically, if a PC can handle the operating system, it should be able to handle Office XP.

For decent performance, you should have at least a 133-megahertz Pentium with 64 megabytes of RAM, though you can get away with 32 megabytes on a Windows 98 system. We recommend at least a 266-megahertz Pentium II with 64 megabytes of RAM. Actually, 128 megabytes gives enough headroom to take full advantage of all components, so you might consider this as the true minimum.

No matter what processor and RAM, the hard drive needs at least 250 megabytes, plus another 50 megabytes of space for each language you intend to use with the international support features. You can check how much drive space you have by right-clicking on your hard-drive icon and selecting **Properties**, as shown in *Figure 2.1*.

If you're planning on doing some cool presentations using PowerPoint, the graphics accelerator should contain at least 8 megabytes of video memory. Optimally, an Accelerated Graphics Port (AGP) card running at times two or faster with 16 megabytes of video RAM will allow you to take full advantage of the animations and effects.

Finally, the PC requires a CD-ROM drive for the installation process.

No matter what processor and RAM, the hard drive needs at least 250 megabytes.

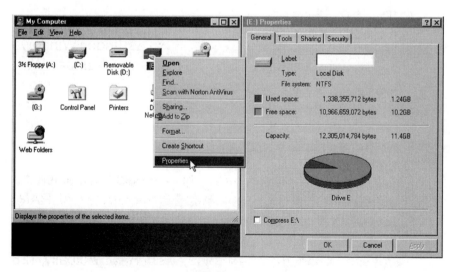

Figure 2.1

The product key is a long series of letters and numbers, separated by dashes.

How to install

Installation of Office XP is relatively easy, and for the most part involves putting the CD-ROM in the drive and letting the computer do the rest. However, a few special circumstances might arise when you attempt to install, depending on the age and type of operating system.

First, no matter which operating system you are using, shut down all other applications. Any application running while installing Office could corrupt the installation.

Pop the Office XP CD into the drive and as part of the installation process it will tell you any running applications that could disrupt the installation process. Then you can turn off only the hazardous programs without much risk.

The other possible glitch some users will run into only occurs on computers running the Windows NT or Windows 2000 operating systems. Due to the secure nature of these systems, users need to be logged on and have full administrative privileges before installing. This probably won't affect a home user, but it might come into play for users on a shared system, or in an office where the security policies are rather strict.

Office XP does some file placement in areas where full administrator access is required.

Near the beginning of the process, users will be asked for the product key. This is a security measure designed to help fight piracy by Microsoft. The product key is a long series of letters and numbers, separated by dashes. A series of fill-in boxes will appear on the screen as shown in *Figure 2.2*. Enter the product key from the

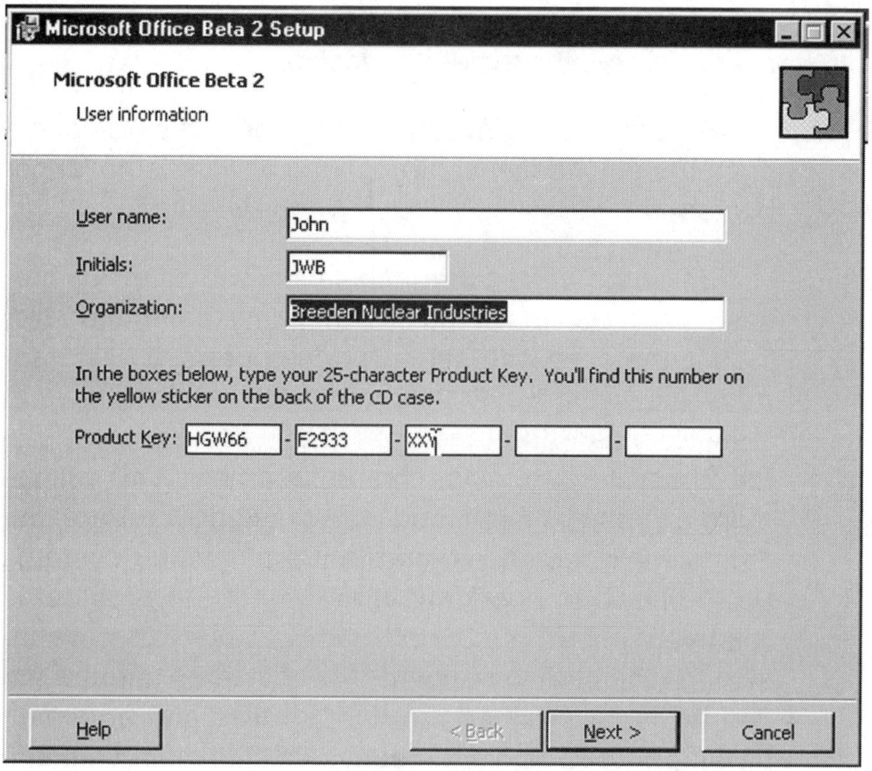

Figure 2.2

main CD case exactly as it appears. If a mistake is made, you will need to reenter the code, but the full mistaken code will remain on the screen, so you can scan and then zoom in on your error.

You can move between fill-in boxes within the code form by pressing **TAB**. In this manner you can move right to the block with the mistake. The code is not case sensitive; you should not have to worry if you use upper- or lower-case letters, no matter what the code on your CD case looks like. If you want to be safe, you can hit the **CAPS LOCK** key, which will capitalize all your letters, yet not affect the numbers.

You can move between fill-in boxes within the code form by pressing TAB.

Keep the code in a safe place in case reinstallation of the program becomes necessary.

The long string of code in the product key is annoying, but until the day comes when software is no longer pirated, it's an evil we will all have to live with.

In some cases the installation will also not precede if certain files in the operating system are too old. This is the case with some NT computers. It also happens with Windows 2000 computers that don't have at least Service Pack 1 installed. If this occurs, users can tell the system to install the needed files. This will require a separate installation process and a reboot, but Office XP does an admirable job at getting systems up to speed to host Office, at least if the problem is relatively minor. Just don't panic. In our tests, we installed Office XP on a lot of older systems and the Office disks always had the needed files and automatically upgraded us. It just requires a little more time and an extra reboot.

Even if you direct the installation program to install on another drive, you will still need some space on C.

One problem that might also be encountered is when you have a system with multiple drive partitions, where the main drive, normally C:, is small and the other partitions are large. This is more common with NT systems, but any system can be configured this way.

The problem is that even if you have a ton of space on other partitions, Office XP is going to want to put a lot of system files on C:, assuming C: is your main drive. Even if you direct the installation program to install on another drive, you will still need some space on C:. The only thing to do is to clear some space on drive C:, and then install the program.

Upgrading from Office 97 Through 2000

If you already have a Microsoft Office suite on your system when you go to install Office XP, you will be asked if you wish to upgrade the older program. This is the default choice and appears in a separate box than the other custom or full install choices as shown in *Figure 2.3*.

It is almost always a good choice to upgrade. Once users get used to all the new features in Office XP, it is doubtful they will want to go back to a previous version. If you do choose the upgrade option, the previous version of Office will be overwritten. Still, due to the large amount of space involved, and because all the old documents will work fine with the new suite, there is really no reason not to upgrade. If you decide to go back later,

Once users get used to all the new features in Office XP, it is doubtful they will want to go back to a previous version.

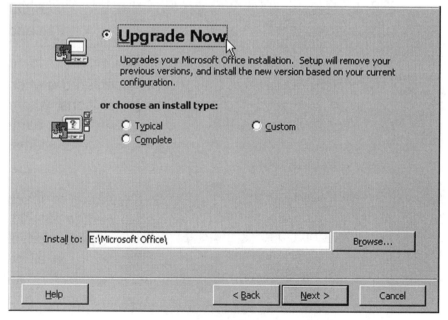

Figure 2.3

just be sure you have the old installation CDs from your previous suite to install later.

If you insist on keeping the old suite on the computer, doing a fresh install accomplishes this so long as you select to keep the old version intact. The program will strongly encourage you not to do this, but you can if you really want. In our tests, both suites existed on one system in harmony, though it was a huge waste of hard-drive space.

Choose the custom installation and select just those components you need.

Typical Installation

What is typical? In terms of Office XP, it depends on what version is being installed. In the past, the Standard Version of a Microsoft Office suite would contain Word, Excel, Outlook, and PowerPoint. The Small Business edition would additionally contain Publisher. Normally the Professional version would add Access to the mix and the highly expensive Premium editions would have FrontPage and a few smaller toolset programs. Check to see which version you have and what is included — preferably before you buy the suite — as you won't be able to install components not included within your version of Office.

No matter which version of the suite you have, if you aren't planning on using all of Office's applications or components, don't use up hard-drive space with a "typical installation." Choose the custom installation instead and select just those components you need.

Why leave the definition of typical up to an engineer who really does not know you or your needs?

Custom and Full How-to

Assuming you aren't upgrading or using the typical installation, the remaining choices are Full and Custom.

When you select the Full option, every component in the Office XP suite edition will be put on the target system. Depending on the way a hard drive is formatted, Office can take 250 megabytes or more. Sure, today's hard drive space is cheap, but why waste system resources?

Depending on the way a hard drive is formatted, Office can take 250 megabytes or more.

The custom installation provides the best option for most users. Users can choose exactly what parts of Office should be installed as shown in *Figure 2.4*.

The custom installation provides four options for each Office application or component. The first is "run from my computer." This is the equivalent of putting the entire pro-

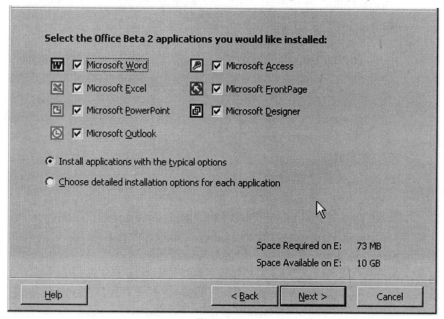

Figure 2.4

gram onto the hard drive. If you pick this option, you won't need the Office CD to use the program once it is installed. For frequently used applications or components, this is the best option.

The second option is "Run from CD." The run-from-CD option puts the front end of the component on the hard drive, but most of the rest of the program is not installed. You will be required to insert the appropriate Office XP CD whenever that application or component is used. This is a good choice if you think you won't be using the program in question too much or if space is limited on your hard disk.

The run-from-CD option puts the front end of the component on the hard drive, but most of the rest of the program is not installed.

The third option is "Do not Install." This option leaves the program out of the suite entirely. This is the best option if you are certain the application or component isn't needed ever. Be sure about this however, as adding an uninstalled application later requires that you begin the installation process all over again at a later date.

The final option for a component is an interesting one. It is "Install on First Use." This is very much like the "Run from CD" option in that only the front end of the component is initially placed on the drive. However, when and if you ever need the component, inserting the Office XP CD will not cause the program to run, but will instead fully install it on your system. From that point on, you won't need the CD to run the application. It is a good option to choose in case you don't think you will ever need a component, but want the option of installing it without much hassle in case you need it later.

If you eventually decide that you have made a mistake and want to put the full version of a previously partially installed or not-installed program on the system, simply insert the install CDs again and you will be given this option as a refresh install.

Although not mandatory, it is good to know that each of the programs in Office contains a set of submodules. The user can install these components of the programs, or the user can block their installation. An example of a module is the spelling dictionary of Word. Of course, it by default is installed, but you can choose to block it, which will keep it out of the computer. If you do this, Word will run fine, but the spell check features of the program will be disabled.

An example of a module is the spelling dictionary of Word.

When you click on the drop-down menu in the installation program, as shown in *Figure 2.5*, you will see a list of subcomponents that are available with the program. A computer icon to the left of the component

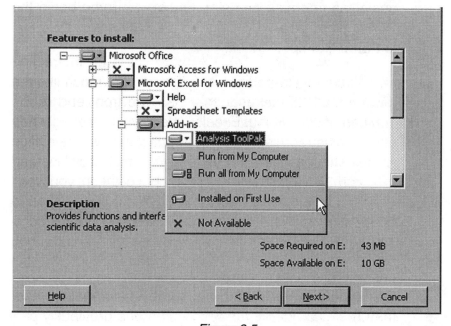

Figure 2.5

Depending on what version of Office you have, you might not have every program available to install.

means the program component will be installed. A red **X** means that it will not be installed. A yellow number one symbol means it will be installed if and when the component is ever used. A sentence listed under **Description** tells you what the subcomponent will do if installed. Left-clicking and holding on the icon will change the status between the two.

Just so you know which programs do what, here is a quick list of what each program does and its proper name. You can refer to this quick list during the install to see if you really need a program or not, or if want to run it off your CD-ROM, etc. Remember, depending on what version of Office you have, you might not have every program available to install.

Word — word processor. This is the most frequently used application. We recommend fully installing it, especially if you plan on using Outlook. If run off the CD-ROM, Word is a little slow when activating advanced features like spell check.

Excel — spreadsheet. Although your job may not require math, Excel acts as a handy organizer of data. You can get by without Excel. Excel can open almost any other mainstream spreadsheet application, too, although some of the coding might be lost. You might consider the "run from CD" option.

Outlook — e-mail, contact database, and schedule. If you plan on using any single part of Outlook, fully install it. If you use a different personal information organizer, then you may not need Outlook. However, Outlook is tightly intertwined with some of the collaboration features of Office.

PowerPoint — presentation package. Totally re-vamped from previous versions, PowerPoint is simple but resource hungry. If you do any presentations, install PowerPoint. If visitors bring presentations, PowerPoint is the most likely application, so it might be a good idea to install just in case, even if it is running from the CD. If you have nothing at all to do with presentations, then don't install this component.

Access — database. A very powerful and advanced database, this version of Access is a lot easier to use than a lot of other databases. Since databases can get very large, it is doubtful that anyone will bring you a data-base. If you don't create databases as part of your job, don't install it.

FrontPage — Web page design and Web site man-agement. Quickly and efficiently make Web pages with-out knowing hypertext markup language, the underlying code of Web pages. Many Web designers use FrontPage, and this version is fairly easy to use. If you have only occasional need of Web pages, you can use Word to cre-ate one, so FrontPage could be skipped.

Many Web designers use FrontPage, and this version is fairly easy to use.

Installing on a Portable Computer

Laptop users have a set of special concerns when installing Office XP. On the one hand, laptop systems generally have smaller hard drives than desktop systems. Also, laptops have the two-fold enemy of a slower sys-tem bus and longer read times from the drives, especially if they are the external type plugged into the main note-book by a cable.

So what you have is a balance between users who don't want to use up all their hard-disk space, but also don't want to have to run programs off the CD-ROM.

The thing to do with a laptop is to install smartly. Only install the Office XP suite programs that will be needed on the road. Use the how-to tips about installing components to do this. If you have a modern laptop with a huge hard drive, 10G or more depending on what applications you run, it's best to err on the side of installing too much, especially if you don't carry your Office CDs with you on the road.

If you have a modern laptop with a huge hard drive, it's best to err on the side of installing too much.

Adding and removing later

Sometimes users will discover the need for a program in the suite that they have not installed, or will grow tired of running a needed component from the CD each time. On rare occasions, space on a drive might become needed and users will have to delete less-used components of the suite to make room for something else.

In all cases, the answer is to do a refresh install. It's pretty easy. Simply insert the first disk of the suite into the drive and wait for the **Autorun** feature to kick in. When it does, select the option to refresh your installation of Office XP. This differs from a new install only in that you can use it to remove programs that are no longer needed.

When you get to the screen where you can choose how you want each component to run, you have the option of installing programs that were previously left out. Also, if you choose not to install a component that is already installed, the component will be deleted.

This process will not affect any of your data files, no matter what program created them. If you totally uninstall a program that created a data file however, you may not be able to open it anymore on that same computer, so be careful. The data file itself, however, remains unharmed.

Uninstalling it All

If getting rid of the entire Office XP suite is what you desire, there is a good way to accomplish this task without need of the original CDs. Using this technique is a bit like taking a baseball bat to a watermelon, in that

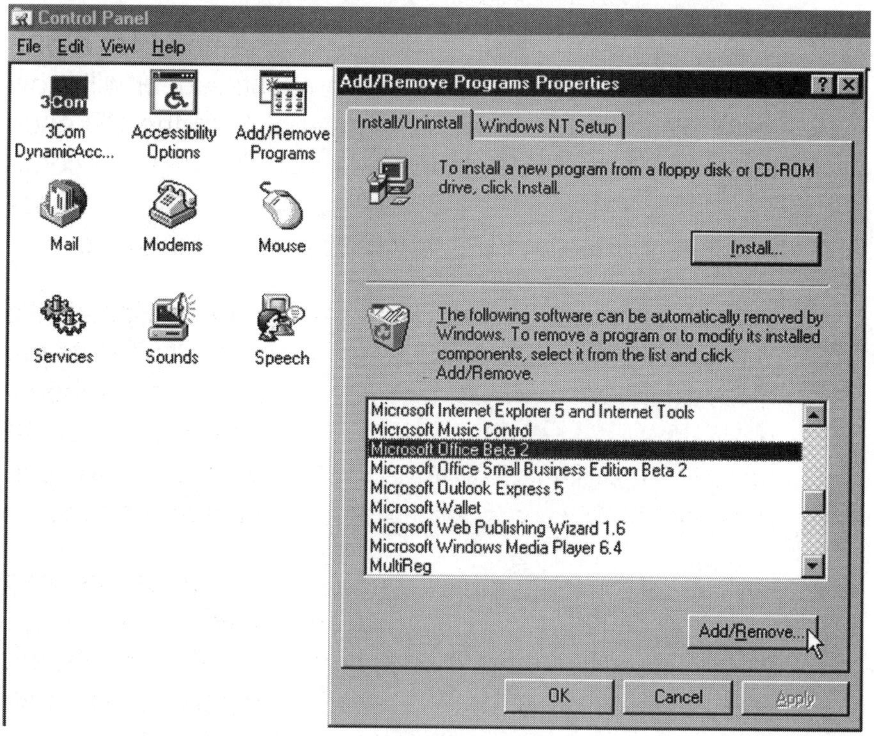

If you totally uninstall a program that created a data file, you may not be able to open it anymore on that same computer, so be careful.

Figure 2.6

You can go back later and reinstall the program if you change your mind.

there will be nothing left. Data files will, as always, remain unaffected.

Simply go to your system's start menu and left-click on it. You should see an option, under the Settings panel, to bring up the control panel. Select the Control Panel option by left-clicking on it. Then, from the Control Panel left-click on the **Add/Remove Programs** icon. Scroll down until you see Office XP listed and select it, as shown in *Figure 2.6*. Click **Add/Remove** and the entire program will be wiped from your system.

You can go back later and reinstall the program if you change your mind, so long as you have the original disks and the copy protection code on the back of the CD-ROM case.

3

Word

Basic Concepts

Microsoft Word is the like the power core for your Office XP spaceship. Almost every computer user needs a word processor. In fact, many users rarely touch anything else. Word even extends itself into Outlook as the default e-mail editor, so even users that never have to type memos or other documents will have contact with Word if they use e-mail.

Word processors began as a way for people to type letters, often for printing and mailing. Microsoft Word really started to extend this concept in the early days to include memos, resumes, envelopes, and eventually e-mail.

Nowadays, there are very few things in the business world that cannot be created using Microsoft Word. From business plans to faxes, bar-coded letters to Web pages, Word has a template that can act as either a foundation for modifications or the actual structure of the document.

Nowadays, there are very few things in the business world that cannot be created using Microsoft Word.

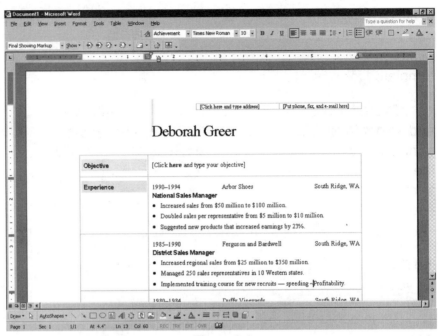

Figure 3.1

The templates are an excellent place to start, but don't be afraid to add a personal touch to your documents.

It is worth noting here that while Word comes with many excellent templates, it often is smart not to use them too often without changes. Remember, potentially millions of people have access to the same templates. We know of a hiring director at a major corporation that can tell at glance which Word template was used for an applicant's resume, such as the one in *Figure 3.1*. And he often throws them away since he is looking for creative employees. The templates are an excellent place to start, but don't be afraid to add a personal touch to your documents.

New in Office XP

More templates, downloadable templates

The most basic concept that will become apparent when using Word is how diverse the templates actually are. This is a dozen-fold increase from previous versions of Office, as shown in *Figure 3.2*, and shows that Microsoft has really listened to the demands of users. Moreover, Microsoft now offers a Web site that includes additional templates for download.

To start Word, simply go to the Windows Start menu and select **New Office Document**. This will bring up the template menu, which is quite extensive. The 15 tabs at the top of the window will give you access to different templates, other than the basic ones that are selected under the **General** tab, which is the default.

If you just want to start from scratch, double-click on the **Blank Document** icon. This will bring up a blank page for you to format in a style you choose.

Be sure that the template you select has a blue "W" in the corner. This indicates that you are opening Word and not some other application in the suite. For example,

Microsoft now offers a Web site that includes additional templates for download.

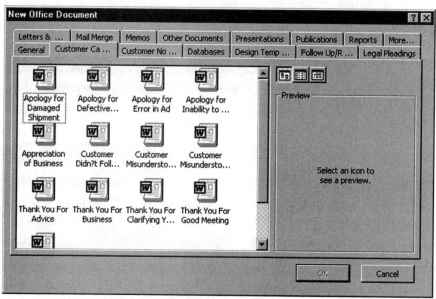

Figure 3.2

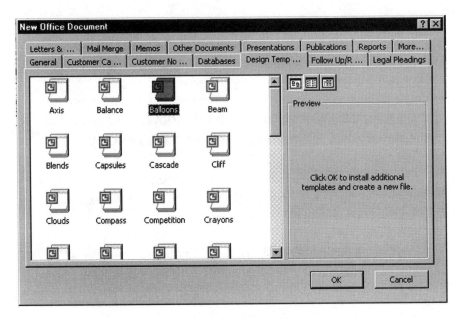

Figure 3.3

Not every template is installed with the standard options when originally putting Office on your PC.

clicking on the **Blank Workbook** icon under the default **General** tab will open Microsoft Excel, the spreadsheet component.

You should also know that not every template is installed with the standard options when originally putting Office on your PC. Some of the more obscure ones have icons, but will require installation. Microsoft obviously did this to save space on users' hard drives.

For example, if you go to the **Design Templates** tab and click the Balloon icon, notice the text in the box to the right says "Click OK to install additional templates and create a new file," as shown in *Figure 3.3*. In this case you will need to have your Office XP CD-ROMs handy. All of the options under the first **General** tab are installed by default.

The Interface

The interface of Word is much easier to navigate than in previous versions. First-time users should have no problem navigating the old serpentine menus of the program. Users accustomed to the somewhat difficult interface of previous versions are also in for a treat.

The most basic change to the interface is the highlighted menus that adorn all aspects of Word. When you place your cursor over almost anything that can be pushed, clicked, or selected, the active area highlights. If you hover long enough, a little box appears with a short description of what that icon, button, or selection does. This should help users learn the interface while they work.

The biggest change to the Word interface is *Smart Tags*. Smart Tags are revolutionary, but they can be confusing if users don't know what is going on.

A Smart Tag is activated as soon as any type of automatic correction occurs. In order to quickly activate this feature, go to Word and type **1** then **/** then **2**. Word automatically changes the typed numbers into the half fraction ½. Hover the cursor over the ½ and a little blue line appears beneath it. Put your cursor on the blue line and a little button appears with a lightning bolt. The lightning bolt is the Smart Tag icon.

New in Office XP

Smart Tags

When you click on the Smart Tag, a menu with three options will appear as shown in *Figure 3.4*. The first option is to **Undo**, or in this particular case **Undo Fraction**. If you select this option, the fraction in this case

> *The interface of Word is much easier to navigate than in previous versions.*

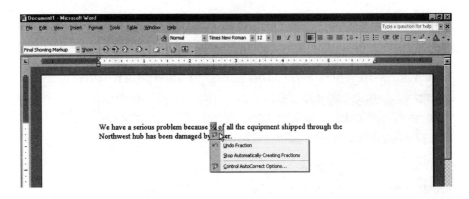

Figure 3.4

*If you find that you have changed something in the AutoCorrect submenu that you regret, you can find it again by clicking on the **T**ools menu and selecting **A**utoCorrect Options....*

will be put back into normal format of 1/2. This will not affect future corrections. If you type 1/2 in a new sentence, Word will again turn your text into a ½ fraction. This is good in case you want one piece of data to be different from the norm.

The second option will be **Stop Correcting** whatever Word just fixed, or in this case to **Stop Automatically Creating Fractions**. If you click this option, the current selection will revert to normal and Word will not correct any future occurrences.

The third selection is **Control AutoCorrect Options....** This will open a menu that gives you access to all automatic correction options. From this window you can add or change all the rules of the program, fully bringing Word under control from one central location.

If you find that you have changed something in the AutoCorrect submenu that you regret, you can find it again by clicking on the **Tools** menu and selecting **AutoCorrect Options....** The **AutoCorrect Options...** command has a picture of a little lightning bolt to its left, so it should be

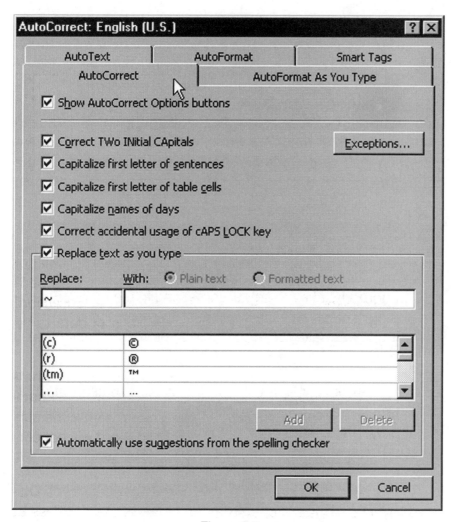

Figure 3.5

pretty easy to locate under **Tools**, or just press **ALT+T** to select the **Tools** menu and then press **A**. The tools options menu should look like *Figure 3.5*.

Now we're going to take a bit of a detour to explain how we represent some commands.

*All of Windows
and all of Office
can be accessed
from just the
keyboard.*

Notice we attempt to show you the most accurate representation of menu or commands by putting those commands and keystrokes in bold. We also add the appropriate underline, just like you see on the screen. So in the case of **Tools**, we have the **T** underlined. Anywhere in Office, press **ALT** and the letter **T** to select the Tools menu. This works with all menus, like pressing **ALT** and **F** for the **File** menu, **ALT** and **V** for the **View** menu and **ALT** and **A** for the **Table** menu. Notice it's the "a" that's underlined—that's how you know which key to hit. To back out, just hit the **ESC**, the so-called "escape" key.

Once you get the menu to pop up, you can press the appropriate key to select something in the menu. For example, pressing **ALT+V** for the **View** menu and then pressing **K** will make the **Task Pane** appear. The Task Pane is a new feature throughout Office XP.

In other places, like any wizard, we also note buttons like **Next >** that you can click or hit the **N** button to make the wizard go to the next step.

All of Windows and all of Office can be accessed from just the keyboard. It just takes a little extra time to figure it out.

◆　　◆　　◆

You can download a free Smart Tags development kit from Microsoft. This lets people and organizations create their own Smart Tags. Microsoft created one as a legal dictionary. Whenever users who have installed the legal Smart Tag expansion type a legal term such as the word "contract," a Smart Tag appears and users can jump to a definition. A few Web pages are starting to put downloadable customized Smart Tag packages on their sites as well, since they work both in Word and within an Internet browser. Users who see a Smart Tag package that interests them are encouraged to download it, since this can truly customize Office to suit individual needs. Customized Smart Tag packages work exactly the same as the standard ones that come with Office. The new packages just contain different information.

The Word Count toolbar can be added to the overall interface with ease.

Users interested in making every word count—literally—now have a powerful new toolbar to help out. The Word Count toolbar can be added to the overall interface with ease. Once installed, the Word Count toolbar analyzes documents in many helpful ways.

To initially activate the toolbar, go to the **View** menu, and then choose **Toolbars**. A submenu appears with 20 selections. Click on **Word Count**.

Alternatively, hover your cursor to the right of where the current toolbars are located. Right-click and the toolbar menu appears. You can select **Word Count** there, too.

New in Office XP

Word count toolbar

This will put a tiny bar on your screen. You can configure the counting engine to look at words, characters including spaces, characters not counting spaces, lines, paragraphs, or pages. Just click the down arrow ▼ next to actual word count totals to see the other statistics. If Word loses track, you can hit the **Recount** button to get a nearly instant automatic total, as shown in *Figure 3.6*.

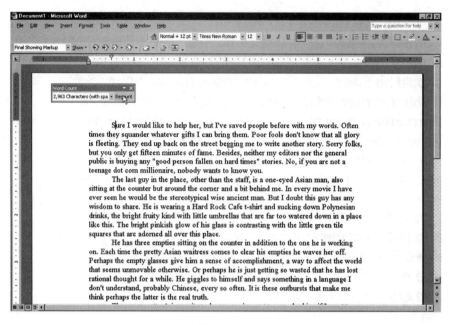

Figure 3.6

Like all menus in Word, the Word Count menu can be moved anywhere on the screen. Simply put the cursor over it and hold down the left mouse button. Then drag the toolbar anywhere. It can be locked into place along the sides, top, or bottom of the screen, or it can hover in any unused place on the main screen.

Skillful management of fonts can make a bland document interesting.

Herding the fonts

Skillful management of fonts can make a bland document interesting. Poor use of fonts can make a "You have just won 1 million dollars" letter appear dull.

The use of fonts is one of the key elements that separate word processors from simple typewriters. Fortunately with Word, changing fonts is relatively intuitive.

The first thing you want to do is to be sure that the **Formatting** toolbar is active. Look for a window on the screen that has a **B**, an **I**, and a **U**. To the left of that there should be the name of the default font and the point size. If you don't see this, right-click on any toolbar and select **Formatting** from the menu. A check will appear, and so will the aforementioned **Formatting** menu.

Be careful, because the menu is quite long, and will need a full screen bar, as shown in the center of *Figure 3.7*. It can't share space with many other toolbars and still be completely on screen, except for very small ones like **Word Count**. If there are little arrows at the end of the toolbar, you will need to left-click and hold and drag the Formatting toolbar to a new area to use all the features.

Alternatively, you can select the **Font...** command. It is the first selection under the **Format** menu. Let's go ahead and open the **Font...** command's window, so press

If there are little arrows at the end of the toolbar, you will need to left click and hold and drag the Formatting toolbar to a new area to use all the features.

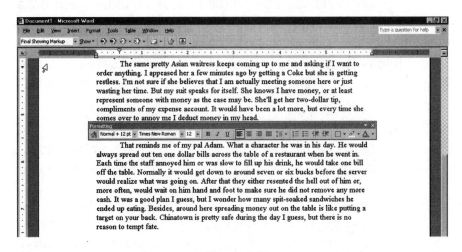

Figure 3.7

If you don't know what you want, you can scroll though the font choices.

ALT+O and then **F**. The toolbar will confirm what you are doing when you make changes.

The first tab under the **Font...** window is the main one you will use to change fonts, coincidentally enough called **Font**. If you know the name of the font you want to use, you can type in the first couple of letters in the **Font:** field window, and Word will automatically go to find what you want. If you don't know what you want, you can scroll though the font choices. The large area at the bottom of the screen shows what the font will look like.

The second panel next to the font name shows any common effects you would like to add, known as **Font Style:**. Here you choose such styling as italics (*italics*), bolding (**bold**), or underline (underline). You can later change these from the font toolbar by simply clicking **B** for **bold**, *I* for *italics*, and U for underline. You can also combine them so text is all three — ***italicized, bold, and underlined***.

The final top-line selection is **Size:** for the size of the font you want. The larger the point size, the larger the font will look. A standard point size for most letters and books is between 10- and 12-point, although text can be bigger or smaller for emphasis or other special events.

Just for your information, 72 points comprise an inch. That means 72-point type is about an inch tall from the top of the tallest character, like a capital I, to the bottom of the lowest character, like a tail of a g.

Beneath these entries in the **Font...** command dialogue, a series of check boxes appear. These are used very much like the options in the second submenu, but are much less common than simple bolding or italics. To activate an effect, click the check box next to the selection you want. Note that not all effects can be used together. For example, **Shado̲w** and **O̲utline** work in tandem, but **En̲grave** won't work on a font that is already **E̲mbossed**. Use the check box font enhancements sparingly, as overuse can detract from the message of the document and can, in some cases, make it difficult to read.

Once you have chosen your font and effects under the first tab window, click on the second tab to modify the font even more. Under the **Cha̲racter Spacing Menu**, you can change the spacing of the font, its position on the page, and at what scale it is represented. Note that spacing in this case refers to the space between the letters, not the space between lines in a document, and thus should only very rarely be changed. In the next section of this chapter we will discuss how to change general spacing within a document.

The final menu under fonts is for **Te̲xt Effects**. These should be used very, very infrequently as each one has the ability to make a document unreadable if used too often. Some examples of text effects in word are **Las Vegas Lights**, which puts blinking colorful dots around words, **Sparkle Text** as shown in *Figure 3.8*, which makes fireworks-like effects occur right in the text, and **Blinking Background**, which makes text seem to highlight on and off.

Once you have made your font selection, the new words you type should follow the format guidelines you have set down. The new settings will affect everything you type following the new settings. It won't go back and change anything you have already typed. It also won't work if you back the cursor up to a previously formatted

Once you have made your font selection, the new words you type should follow the format guidelines you have set down.

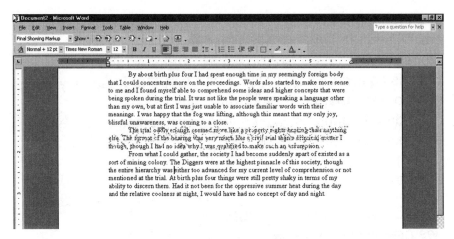

Figure 3.8

piece of text and type something new. This is because what is really happening is that Word is putting a hidden flag at the point your cursor is sitting when you go to the menu that tells the program to "change everything past this point." Hidden codes can sometimes be annoying, but we will tell you how to find and root them out later on in this chapter. The ability to find and change the codes manually is a huge improvement over previous versions of Office.

You can go back and change previously formatted text. What you need to do is to highlight the text you want changed by drag selecting the targeted words, sentences, or pages. To change the whole document, just type **CTRL+A**, which means "select all." Then you bring up the format menu with the text still highlighted. When you make changes, it will be applied to the highlighted area.

The ability to find and change the codes manually is a huge improvement over previous versions of Office.

Fit formatting

The computer's ability to precisely format documents in a way to best get the message across is the other reason word processors eventually killed typewriters. There are millions of format codes for most word processors that can make documents look different ways. Word is no exception. But like a parched castaway adrift in an ocean, there is sometimes water everywhere but not a drop to drink. Or more precisely, there is often not a drop of the right kind of water. Too much formatting or uncontrolled formatting is often worse than simply creating a plain document.

To help users rustle up the right formatting, Microsoft has greatly improved the control interface for formatting documents.

The best example of more user control is in the **Styles and Formatting**… Task Pane. To activate this pane, go to the **Format** menu and select the **Styles and Formatting**… command. By default, a window will open to the right side of the screen. This **Styles and Formatting…** menu will show all the style choices available to a user. Near the bottom of the pane, an option bar appears next to the word **Show:** that lets users select what information to display in the pane. A good choice is often to have the pane display just the **Formatting in use**. This way you can quickly select previously defined formats by simply left-clicking on them and change the format while ensuring that the document keeps a consistent look.

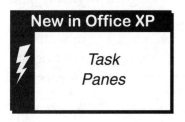

Each kind of template in Word contains a certain style. The style of text varies from the template selected. The Blank Document template selected earlier has very

Too much formatting or uncontrolled formatting is often worse than simply creating a plain document.

basic text, indented text and some heading fonts. Other templates contain other fonts. Of course, you can add fonts to whatever document you're working in.

In addition, the **Styles and Formatting...** pane, shown in its default location along the right of *Figure 3.9*, is helpful because if you hover over a style, you can left-click to bring up options regarding everything in the current document set to that style. The most helpful use for this is to first choose **Select All X Instance(s)**, where **X** is the number of times that format was used. Then you can

New in Office XP
Select all instances of a particular style within

choose **Modify Style...** to change all occurrences in the document at the same time. This is most helpful if you

You can add fonts to whatever document you're working in.

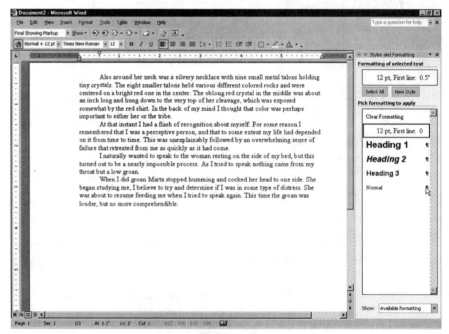

Figure 3.9

discover that certain elements of your document, such as special section-break headers, need to be larger or smaller. Use the **Format** button in the bottom right corner of the **Modify Style...** dialogue to make the changes.

Another good use of the formatting Task Pane is the **Clear Formatting** option. First, select the type of format you want to clear using the **Select All X Instance(s)** command. Once selected, click on **Clear Formatting** and all occurrences will be reset back to the template's defaults. This will not affect any hyperlinks that you may have entered into the program, either as straight text or live links.

You may not know it, but you can now also manually select different parts of a document for formatting by hand, even if the document sections are separated by text that you don't want to modify. To do this, highlight any text by left-clicking on it and dragging your highlight over the targeted area. When this is done, hold down the **CTRL** key and move your cursor to other areas that also need selected. Drag-select those areas in the same way, never letting up on the **CTRL** key. What you have highlighted will be selected, but everything in between will not, as shown in *Figure 3.10*. You can then release the **CTRL** key and change the format of the selected text.

New in Office XP

Select or highlight no consecutive text

You can now also manually select different parts of a document for formatting by hand.

The Task Panes can help format entire documents at one time, but it is often easier to just set down formatting rules when you begin a document, especially if you don't foresee much change other than bolding and simple

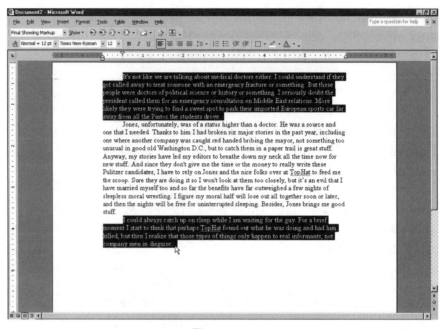

Figure 3.10

changes that can be made with one button. Since everything after you make changes will follow your rules, you should set up simple documents as soon as you load a blank page.

To do this, go to the **Format** menu and select **Paragraph**. The first option will be the **Indents and Spacing** tab. You can change the general alignment of the document from here, though by default it is set to the commonly used left justification. Here is also where you control line spacing. You can set the spacing between lines to anything, from single and double space to almost any amount.

If you need to set the spacing to something not very common, go to the **Line Spacing:** area and choose **Multiple** from the drop-down menu. You can put 3 lines

You can set the spacing between lines to anything, from single and double space to almost any amount.

or 3.5 lines. To be a bit more precise, select **Exactly** and choose the amount of points between the lines. This is measured from the baseline of each text line (the baseline is the invisible line on which all characters line up).

You can also ensure consistency by going to the **Format** main menu and selecting **AutoFormat....** This will prompt Word to search the entire document for instances of formatting that are not consistent with the rest of the document. This is sort of done automatically as you type with the Smart Tags, but you can force the automatic format checker from this menu as well. There are two options when using this format detective. The first is to run the checker and let it make changes. This is the **AutoFormat now** option. The second—and more advisable—option is **AutoFormat and review each change**. This lets the user approve any changes that are made. Office XP is good at what it does, but like all software, it is not perfect. It is good to always check the "automated" work.

Office XP is good at what it does, but like all software, it is not perfect.

The most powerful formatting feature, however, is the ability to reveal all the formatting codes within a document. This is done via another Task Pane that can be activated by going first to the **Format** menu and then selecting **Reveal Formatting**. The default location for this menu is on the right side of the screen, and because it is quite large it will overwrite any format panes you already have in that area. Otherwise, the viewable area of the document would become almost too small to be functional.

New in Office XP

Reveal Formatting command

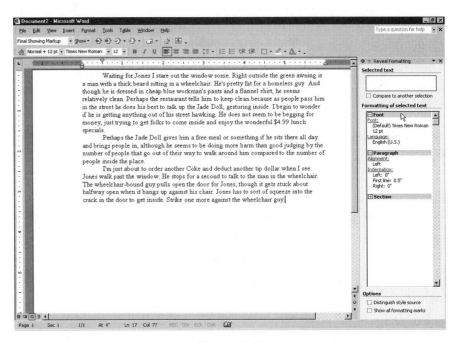

Figure 3.11

The first function of the **Re<u>v</u>eal Formatting** pane is that it will tell you at a glance how the current selection is configured, as shown in *Figure 3.11*. It will display the font type and point size, the language the selection is in, the al<u>i</u>gnment, the indentation, and the spacing.

Near the bottom of this pane is an unchecked box that says **Show all formatting marks**. Clicking this will show all the hidden marks and format codes in the document, like all the dots where spaces are located, showing everything that isn't hidden. Compare *Figure 3.12* to Figure 3.11 to see the difference. Figure 3.12 is identical, but has all formatting codes revealed. Clicking the box again will make the codes invisible once again. Some people love the codes, but many also find them confusing. You can see these codes anytime by pressing the paragraph

Some people love the codes, but many also find them confusing.

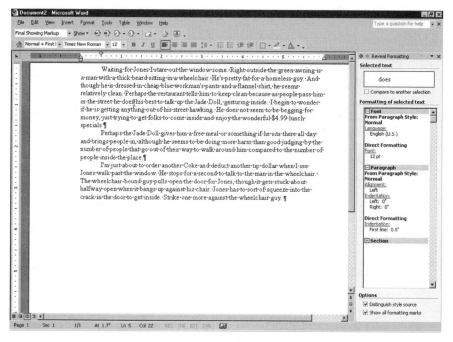

Figure 3.12

symbol ¶ on the **Standard** toolbar, by default appearing at the top of Word's toolbars just beneath the menus.

But we think the nicest feature is the pane's ability to compare two parts of a document for formatting differences. This way you can tell if some part of the document is actually slightly larger, indented differently, aligned slightly off — or if your eyesight is just getting bad from too many hours in front of a monitor.

The nicest feature is the pane's ability to compare two parts of a document for formatting differences.

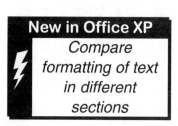

New in Office XP

Compare formatting of text in different sections

To activate this feature, make sure the **Reveal Formatting** pane is open. Go from the text you want to compare to the new text and drag-select it. Then, click the **Compare to another selection** box

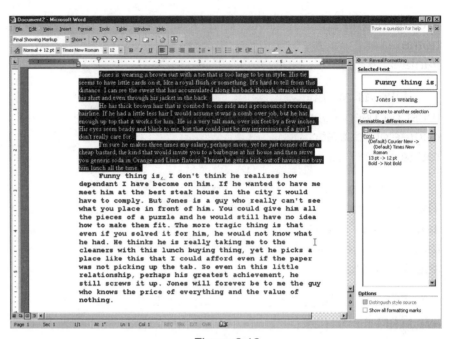

Figure 3.13

near the top of the pane. Go to the new selection and highlight it. The large pane window will either say "No formatting differences" or tell you exactly how the two differ, as shown in *Figure 3.13*. You can continue to select more text to compare to the first by simply highlighting new text. You only have to change this pattern if you want to begin comparing text to a new selection. In this case, click off the check box and select the new comparison text.

A splash of color

Adding a bit of color here and there to a document can be both beneficial to the reader in terms of readability, and to the author as a means to push certain points. And reading black text on a white background is eventually going to get boring anyway. Besides, color printers are becoming commonplace in today's office and most home users use a color inkjet printer. Why not take advantage of color?

Like all methods of sprucing up a document, color must be used sparingly. One also needs to have some degree of aesthetic talent, at least to know things like thin red text is difficult to read on a green background.

There are two main ways to add color to documents. The first is to add color to the background itself, and the second is to change the actual color of the text. If you really want, you can do both at the same time for an extra special effect.

The Formatting toolbar has an entire line on your screen to itself.

First, you should have the **Formatting** toolbar active. That is the one that you previously used to change fonts. Go ahead and left-click anywhere on an existing toolbar to bring up the active menu. If **Formatting** is not checked, select it. The other thing that you need to be careful with is that the Formatting toolbar has an entire line on your screen to itself. This can be locked in a corner or floating anywhere by itself.

The reason you need to have access to the entire toolbar is that you will need to use some icons all the way on the right side of the bar to add some color to the document.

For this exercise, we will be concerned with two icons on the toolbar. The first is represented by the letter A sitting on top of a colored bar, like so: <u>A</u>. The other looks like a pen slanted to the right, also sitting on top of a colored bar. For both icons, the color of the bar represents the color that each will produce when used. When you start, the bar under the highlight or pen-shaped icon

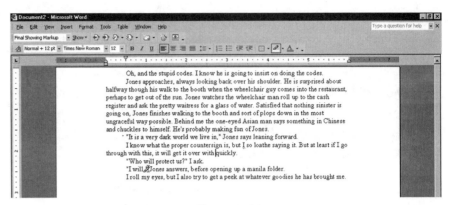

Figure 3.14

should be white and the text-color icon bar, sitting under the A, should be black. This will produce back text on a white background.

The highlight icon is an active or hot icon, while the text icon is passive. That means that to work with the highlight icon, you will need to actually select the icon and then move the cursor around the document. The text icon, once changed, will change all the text typed from that point forward.

To use the highlight feature, click on the little upside-down triangle to the right of the icon. A group of colored squares will appear. Select the color you wish to use to highlight your text. Once selected, go back and click on the pen itself. When you move the cursor back to the document, it will look like a pen. If it does not, you need to click on the highlight icon again.

Use the virtual pen just like you would a highlighter on a piece of paper, as shown in *Figure 3.14*. Simply left-click and drag the pen across the words you would like highlighted. The background will be changed to the color you chose, though the text itself will remain the same.

To work with the highlight icon, you will need to actually select the icon and then move the cursor around the document.

If you make a mistake, you can go to the edit menu and select **Undo Highlight** or you can go back to the triangle beside the pen on the toolbar and select **None**. Once **None** is selected, dragging the highlighter across a colored area will erase all color, turning it back to white.

To actually change the color of the text, you can use the little triangle beside the **A** icon in much the same way. Just select a box from the 40 colors or choose **More Colors…**. All future text you type will come out in that color.

To change the color of text previously typed, select the area using your cursor and with the area still selected, go up and choose a new text color. If you make a mistake, you can go back, select the mistaken area, and then choose black.

If you wish to change all the background for an entire document, this can be done by going to the **Format** menu, selecting **Background** and choosing a color or even a watermark. As a warning though, this will change the entire document, and may conflict with previously set highlight or text colors.

Word can also insert a watermark for you if you want to protect your printed documents from theft, or if you wish to distinguish draft copies from final ones. The watermark will appear when the document is printed. It could even be a vanity-type mark, like a company logo.

To place a watermark, simply go to the **Format** menu, then the **Background** menu and select the bottom option: **Printed Watermark…**. You can then set the placement of the watermark as well as type what the watermark will say, or insert a photo from disk. Placing a

Word can also insert a watermark for you if you want to protect your printed documents from theft.

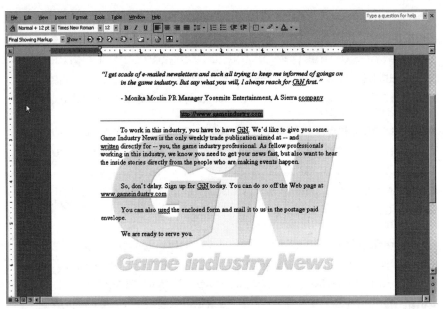

Figure 3.15

watermark will not affect the on-line document, but will be evident when printed. *Figure 3.15* shows what a document with a slightly faded watermark will look like.

Word gives users the opportunity to add actual pieces of art to their creations.

Artistic license

In addition to just coloring text in a variety of ways, Word also gives users the opportunity to add actual pieces of art to their creations. This can be done in several ways.

If you have a photo on a disk that you would like to add to your document, the quickest way to accomplish this is to activate the **Picture** toolbar.

Once active, the most important button on the toolbar is the icon all the way to the left. It is colored and looks like a sun rising over the mountains. Clicking on this will open another window that looks exactly like the file menu in Windows. Simply locate the file you wish to import, say one on your floppy disk in the A: drive, and click on it. The picture will be inserted into the document in the spot where the cursor is sitting. You will get a preview of each picture you can use, as shown in *Figure 3.16*.

Once the photo is in place, you have access to basic editing tools from the same toolbar, which will be grayed out unless a photo is selected. Make sure you click on the image you wish to modify before you mess with the editing tools.

From the left you have **Color**, **More Contrast**, **Less Contrast**, **More Brightness**, and **Less Brightness**. The **Color** button can be used if you want to, for example, turn a color image into a grayscale image. It can't, however, turn black-and-white or grayscale images to color, since the color information is either not included with the picture or never included to begin with. Clicking on any of

The picture will be inserted into the document in the spot where the cursor is sitting.

Figure 3.16

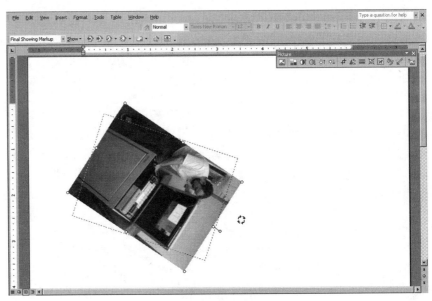

Figure 3.17

the more or less buttons will have an effect on the picture, though you may have to click it multiple times for any real change to occur.

Some of the more advanced tools, separated from the basic ones on the toolbar by a line, are **Crop**, **Rotate**, **Line Style**, **Compress Pictures**, and **Text Wrap**. Cropping selects an area within the photo to display, eliminating anything outside that selection. Rotating spins the photo—sideways, upside down, or anywhere in between. To rotate a photo, click **Rotate** and a series of dots will appear around the photo with one green one. The green one is the "hot dot." Grabbing it will let you turn the photo around any of the 360 degrees as shown in *Figure 3-17.*

Line Style puts a border around the image.

The **Compress Pictures** icon can reduce the "size" of an image. Every time you add a picture to a document, the file size gets larger and larger. Let's say you add a high-resolution 8x10 picture to a document, but in the final document, it will be about 3x5 inches in size. Pressing this button reduces the size so that the file will be 2 megabytes instead of 10 megabytes.

New in Office XP

Compress Pictures option

Also, some documents are never really meant to be printed out. For a photo to look good when you print the document out, that image needs to be at least 200 dots per inch or dpi, which is the default next to the **Print** selection in the **Compress Pictures** dialogue. But let's say you're going to make a Web page. When you click the **Compress Pictures** icon, you can select the **Web/Screen** radio button, which will reduce the dpi of the images in the document to 96 dpi.

The final important icon on the pane is **Text Wrapping**. This is helpful because the photo you imported does not have to stay in one place. You can grab it and drop it off anywhere in the document, even over the top of text. The **Text Wrapping** icon will set up rules for how you want the text to behave around the photograph, whether it should run straight though, form a box around the image and remain outside of that perimeter, or tightly move up as close to the photograph as possible, as shown in *Figure 3.18*.

Users don't always have a photograph readily available to insert into documents, only an idea. When this happens, you need to go about adding art in a slightly different way. Go to the **Insert** menu and then select the **Picture** command and then click on **Clip Art...**. This will open the **Insert Clip Art** Task Pane along the right side of the screen.

Users don't always have a photograph readily available to insert into documents, only an idea.

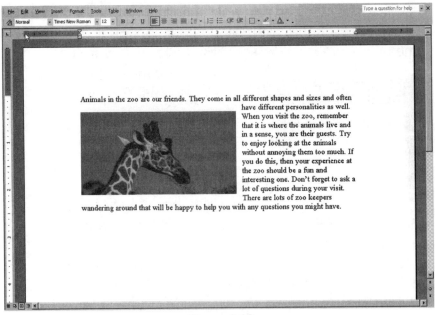

Figure 3.18

Word comes with a nice variety of clip art. You can type in search text in the first box and see if the Office collection of clip art has what you need. For example, when you type "hammer" and push the **Search** button, you will get several examples. If you're connected to the Internet, Office even automatically checks the Design Gallery Live Web site at Microsoft.com.

See an image of a hammer you like? Simply click and drag it into the document or click on the image to bring up your selections on what to do with the image.

Once there, you can resize it by holding the cursor near the edge of the art, it will turn into an arrow, and holding down the left mouse button. Moving the cursor will then change the size of the picture. Enlarging clip art too much, however, will make it appear blocky, so only slight changes are advisable.

If you're connected to the Internet, Office checks the Design Gallery Web site at Microsoft.com.

Microsoft maintains a huge library of clip art files that can be grabbed from Word with ease. Even if you find something that works included with Office, it is often prudent to check on-line, too, as there might be something better in the larger on-line library.

New in Office XP

Automatic clip art search on Design Gallery Live Web site at Microsoft.com

The other type of art that you can add is not a picture at all, but something called WordArt. WordArt is just a fancy name for a picture of a sentence or word.

Go to the **Insert** menu again and select **Picture** and then **WordArt...**. You will see a series of colorful WordArt designs. When you click one you will be brought to a dialogue where you can change "Your Text Here" to whatever you want the art to say. You can also change the point size of the text. Clicking **OK** puts the art into the document. At that point it acts like just another piece of art, and you can change its position and properties, such as word wrap or size, at will.

Table talk

A powerful way to display information in Word, and one that is infrequently used, is to create a table. Think of a table as a mini-spreadsheet. In fact, Excel spreadsheets can be inserted into Word as tables if you desire.

If you have a ready-made table in Excel, all you have to do is to insert it.

If you have a ready-made table in Excel, all you have to do is to insert it. Open both the Word document you want to use and the Excel file with the information. Highlight the information in the Excel worksheet that you want to use and go to the **Edit** menu and click **Copy**, or just press **CRTL+C**. Then switch to the Word document and put the cursor where you want the table to appear. Go to the **Edit** menu and click **Paste**, or just press **CRTL+V**. The information will be inserted as a table.

> *To create a table from scratch is not very difficult.*

To create a table from scratch is not very difficult either. In Word, put the cursor where you generally want the table and go to the **Table** menu and select **Insert** and **Table**. A menu will pop up asking you the dimensions of the table. Set the number of columns and rows you want, and click **OK**. It is also sometimes a good idea to click on the box that says **AutoFit to contents**. This will make the table initially small, but with cells that will expand as you type.

Simply select the first cell in the column and type your field. Then press **TAB** to move on to the next field. If you have selected the AutoFit option, you may run into slightly uneven fields, although they will be as large as the largest entry and should look fine on the paper.

To spruce up your tables somewhat, activate the **Tables and Borders** toolbar. With your table selected, push the icon that has a burgundy-topped table with a lightning bolt called **Table AutoFormat**. This will give you a lot of options for making your bland table look better. Choose a different color scheme or perhaps a slightly different layout. When you click **Apply**, the new format will be applied to the existing table. With this **Tables and Borders** toolbar you can also change the way the table is aligned on the page. The default is to align it along the top left, but you can also choose center, right alignment, top or bottom. Experiment with this function to see what looks best for your specific file.

Automatic outlines

Sometimes nothing shows the flow of ideas better than an outline. But in the past, creating an outline required users to manually set them up, putting in all the tabs themselves and using proper outline numbering systems. This could often lead to confusion.

Word does most of the work. All you have to do is activate the outline function and supply the data.

For simple numbering or bulleted lists — the simplest form of outline — click on either the **Numbering** icon (with the numbers 1 2 3 on it) on the Formatting toolbar. For bullets, click the **Bullets** icon (with three blocks lined up).

When activated, each time you hit return you will be given either a new bullet or a new sequential number. When you are finished with the list, click on the same activation button again. A highlight will turn off and you will be returned to normal use patterns within Word.

More complex outlines, like the numbers followed by indented letters followed by more indented numbers pattern, require a bit more work to set up, but not much more.

Place the cursor where you want an outline to start. Go to the **Format** menu and choose **Bullets and Numbering...** You will find several tabs available and several outline examples under each one. Select the one that you want to use. The most complex ones are located under the third tab.

At this point you can start typing your outline. Notice that when you hit return, the next item is numbered on the same level as the last one. To move deeper into the outline, like to switch from letters to numbers, go to the

In the past, creating an outline required users to manually set them up, putting in all the tabs themselves and using proper outline numbering systems.

Formatting toolbar and click on the **Increase Indent** button. The **Increase Indent** tab is to the right of the **Numbering** and **Bullets** button.

To assign a higher level of number or letter to a line, simply hit the **Decrease Indent** button. In this manner you can easily create an outline without much worry about messing up the format.

The majority of documents created with Word are still bound for a printer.

Putting it on paper

Despite the emphasis on the Web and on-line sharing of documents, the majority of documents created with Word are still bound for a printer. Thankfully, with the new Word this is no longer a guessing game.

When you need to know exactly how your document will look when printed, go to the **File** menu and select **Print Preview**, or in the **Standard** toolbar, select the icon that has a blank page of paper with a magnifying glass on it. This will take you to a separate screen where, by default, you will be viewing pages at 55 percent size, as shown in *Figure 3.19*. This is great for getting a good look at the different pages in the document for any obvious errors. You can also check to see if there are any minor mistakes, like section heads being at the bottom of a page and the section starting on the following page.

New in Office XP

More powerful print preview

For really large documents, there is a button in this **Print Preview** window that looks like four pieces of paper sitting on a darkened field called **Multiple Pages**. When

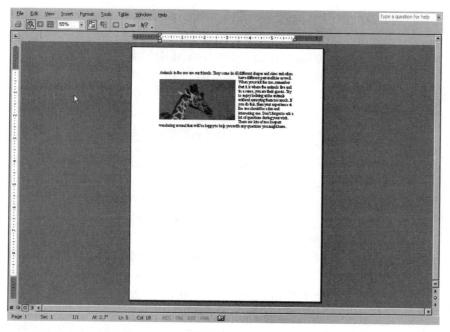

Figure 3.19

clicked, you have the option of putting up to six pages on the display at once. You won't be able to read the text at this level unless it is very large, but graphics and tables will be apparent. This view can be used to quickly see how large documents will look when printed.

When you are finished with this view, click on the **Close** button to return to normal. You can actually work in this view if you increase the magnification to normal, but it is not a good idea because automatic functions like spell check will be disabled. You can tell if you are in this special view because it will say **(Preview)** at the top of the screen next to the title of the document. Be sure to click **Close** before going back to do any last-minute editing.

If you like what you see, however, you can click on the **Print** icon that looks like a printer on the far right side.

You have the option of putting up to six pages on the display at once.

It's normally best to close this view and then go to the **File** menu and select **Print...**. Advanced printing options, like adjusting paper size, are found on a separate submenu called **Page Setup...** under the **File** menu.

Advancing to the head of the class

Although this book is aimed mostly at the special features found in Word to increase the productivity of users, there are a few features that, although helpful, probably will be used by only a few folks.

The most important of these features is collaborative editing and sharing of documents. This feature is so important to office users — and worthless to single users — that we separated it out into an entire chapter. If you want to collaborate with Word or any program in the Office XP suite, you will want to read Chapter 13: All Together Now. Otherwise, you can instead concentrate on more single-user tasks.

One useful advanced feature in Word is its ability to translate documents from one language to another. To translate text, go to the **Tools** menu and then select **Language** followed by **Translate...**. This will bring up the **Translate** Task Pane. It's another big pane on the right of the screen.

> *One useful advanced feature in Word is its ability to translate documents from one language to another.*

New in Office XP

Translation tools

To select words to translate, simply highlight the chosen phrase from your document and then select Current Selection from the radio button menu or the entire document button if you choose. Then select the language translation you want done, like taking **English (U.S.) to Spanish (Spain-Modern Sort)**. Press the **Go** button.

The completed translation will appear in the **Results** window for viewing. If you want to replace what was selected with the new language, choose the **Replace** button. This will erase the former language and put the new one in its place.

You can also type selections into the **Text:** field at the top of the Task Pane for translation, in case you need to know the French word for "cat."

Word can only translate to languages that have been installed. If you remember from the installation chapter, each language requires about 50 megabytes of hard-drive space. If you find you want a language that you do not have, you can do a refresh install of Office by simply inserting the main CD-ROM into the PC again. Be sure that all Office applications and just about everything else on your system is closed when you do this.

Each language requires about 50 megabytes of hard-drive space.

4

Excel

Excel exhibits a dual personality. The uninitiated might think this powerful application would only serve the cosmos of accountants and statisticians. But those of us who have launched this spreadsheet application have discovered almost infinite possibilities.

Excel provides excellent mathematical tools for accountants, statisticians, or anyone who works with numbers. But Excel also endows the not so numerically inclined with an organizational tool unlike any other:

♦ Tap into relational databases.

♦ Get and automatically update data from the Web.

♦ Generate sophisticated — and attractive — charts, graphs and tables.

♦ Create basic analysis of any data without having to remember your first algebra class.

Excel boosts almost anyone into a stratosphere where your boss will be nothing less than impressed.

Excel endows the not so numerically inclined with an organizational tool unlike any other.

Basic Concepts: Spreadsheets 101

Each of these boxes is considered a cell and each has an address.

Launch Excel and enter a world of boxes and little rectangles. As shown in *Figure 4.1*, Excel pops up with a grid of rectangles on the screen. Think of the boxes as a place where you can store things — numbers (of course), formulas, words, and even art or colors. Each of these boxes is considered a cell and each has an address, as indicated along the left side with numbers and along the top with letters.

Click on any cell and the address appears in the top left just above the top left cells. Click on any of the letters to select an entire column and on any numbers

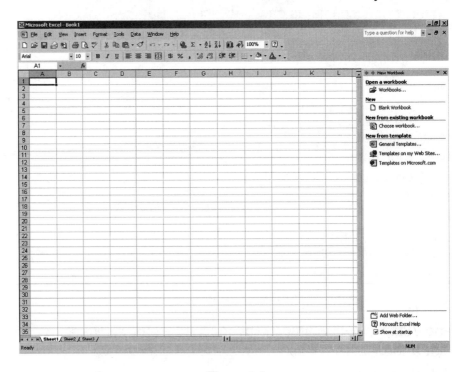

Figure 4.1

for the entire row. Hover the cursor over any cell, press and hold your left mouse button and then drag to select a number of cells.

Users of Office 97 and 95 will notice the new subtle shading now, so everything remains visible when selecting a single cell or a group. Experienced users may also notice the new Task Pane along the right side. The Task Pane offers some of the basic options like access to recently opened files, create new files, opening templates and more.

Users of Office 97 and 95 will notice the new subtle shading.

For the time being, ignore the Task Pane and close it (click the **X** on the top right).

Three different contents are normally entered into a cell:

1. Text
2. Numbers
3. Formulas

Select the top left cell (that's cell address A1) and type your name, then hit **Enter**. Now type a number — let's say **123** — and hit **Enter**. Follow up with **456** then **Enter** and **789** then **Enter**. Excel figures out what is a number and what is text.

New in Office XP

Task Pane

Now, here is a basic formula. Let's add 123, 456, and 789. Hit the = key (equal marks). That clues Excel that a formula is about to be typed in. Move the cursor and click on the cell with 123 inside it. A marquee appears around the numbers and the cell address appears beside the equal marks.

Now, hit **+** (the plus mark key) and click on the cell with 456 inside it. Now the marquee appears around 456 and that cell address appears as part of the formula. Hit **+** (the plus mark key) again and click on the cell with 789.

Excel XP helps the user keep track of what's where.

Excel XP helps the user keep track of what's where. Notice how each cell has a different color around it and each cell address has the same color (*Figure 4.2*). This helps users track where formulas relate to one another.

Now hit **Enter** and the formula **= A2+A3+A4** disappears, replaced by 1398 — the sum of 123 plus 456 plus 789. If you change any of the numbers in cells A2, A3, and A4 the total changes instantly.

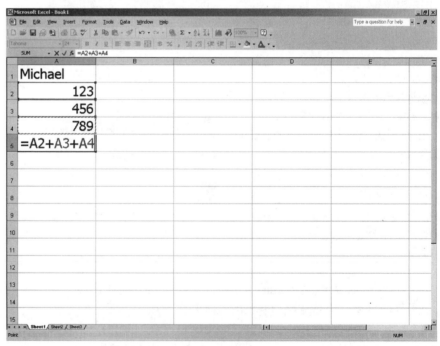

Figure 4.2

Excel can use some entries that, on the surface, do not look like a number at all. Go to cell C1 and type a date— in this case, "November 7" — and hit **Enter**. Under default settings, the entry will appear to change to "7-Nov"; just ignore that. Now type "11/14" and hit **Enter**. The entry changes to "14-Nov" in Excel. Let's type "11-21" then hit **Enter** and watch "21-Nov" replace your entry. Finally, type "Nov 28" followed by **Enter** and see "28-Nov" appear. If "28-Nov" did not appear but "Nov. 28" stays put, click on the cell and type it again without the period after the "Nov" abbreviation. To edit a cell without retyping the contents, hit **F2**.

Excel understood — no matter how it was typed — that the entries were all dates. But Excel also understands these dates as numbers that can be used in a formula. Notice how each date is a week after the previous seven days.

Press = (equal marks key) and click on cell C4 with the contents "28-Nov" inside it. Hit the + (plus mark key) and then **7** and **Enter**. The entry appears as "5-Dec"— seven days after November 28.

Click on the C5 cell with "5-Dec" inside it. Notice the heavy lines around the sides of the highlighted cell and a square in the bottom left corner. As the cursor hovers over the bottom left corner, notice the large plus cursor turns into a thin plus sign. Left-click down and hold, dragging the box down three cells. Notice that "12-Dec," "19-Dec," and "26-Dec" all appear. Each is seven days later!

Experienced users know Excel just figured out that each is seven plus the cell just above it. The formula from the C5 cell is relatively adjusted so cell C6 contains the

Excel can use some entries that, on the surface, do not look like a number at all.

formula "=C5+7" and cell C7 contains "=C6+7," and cell C8 contains "=C7+7."

In Excel 2002, after this dragging, a Smart Tag appears. It's the icon with the lightning bolt on it. Click on it. Excel now checks to see if the user wanted the relative filling or if the user just wanted to copy the formatting.

Sometimes numbers need to be entered as text.

New in Office XP

Smart Tags

Speaking of formatting, the dates look a little weird. Highlight all of the cells that contain dates, from C1 to C8. Now right-click the mouse and a menu drops down. Select the **Format Cells...** item.

All the different format options appear, but for now, in the left box, click **Date**. Several options appear in the right box, so pick the third option and click **OK** (*Figure 4.3*).

Now all dates appear in a standard number format. On my screen, the dates 11/07/00 to 12/26/00 showed up. That's because this chapter was written during the year 2000. Whatever year you're in will be the automatic year Excel assumes you mean, so you might see 11/07/01 to 12/26/01, or 11/07/02 to 12/26/02.

Sometimes numbers need to be entered as text. Click on cell E1. Let's pretend you're in charge of your community's spring May Day festival, which some over-achieving marketing person decided to call 1MAY. Type in "1MAY" and hit **Enter**. Notice, like the previous dates above, Excel assumes the entry is a number and translates it to the default "1-May." It just needs to be a text entry exactly as typed.

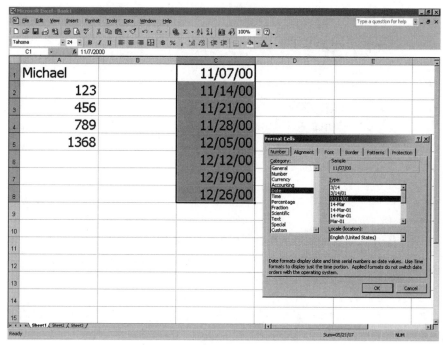

Figure 4.3

The single quote lets Excel know the contents will be text.

In cell E2, type ' (single quotation mark) then type "1MAY" and hit **Enter**. This time, Excel left it alone, just as typed, although the single quote mark disappeared. Just like the equal mark alerts Excel the cell contents are a formula, the single quote lets Excel know the contents will be text.

Now, before we launch into some heavy-duty information, novice users should play around with Excel. Up until now, we've just dealt with the basics. Take Excel for a test drive, entering different information into its cells, dragging to fill the contents, trying out more complex formulas.

The Interface:
"What are you doing, Hal?"

Microsoft went leaps and bounds to make the different applications within the Office suite look similar. Excel looks a lot like Word — at least on the surface. Spend a little more time with Excel; it will look more and more different.

The default appearance includes the menu items across the top: **File**, **Edit**, **View**, **Insert**, **Format**, **Tools**, **Data**, **Window**, and **Help**. You'll find all menu selections are the same across most Office applications except for the **Data** menu item.

Click on any of the menus and the drop-down list might seem a little short. That's because Microsoft thinks users are overwhelmed by too many choices. Menus list only the most frequently used commands. Pause for a few seconds without selecting a command and the rest of the menu appears. Or, if you can't wait, click the little arrows at the bottom of the dropdown menu to show the rest.

Except for a couple of new ones, most are from previous versions of Excel. Microsoft made improvements in existing commands, as already mentioned some, and others will be discussed a little later.

Below the menu items are probably two rows of icons to quickly access some commands. These sets of toolbars are standard, but Excel offers many more. Hover the cursor off to the right of the existing toolbars and right-click with the mouse. A drop-down menu offers 18 toolbars for various uses, including charts, pictures, and pivot tables. At the bottom of the list is the word **Customize....** Toolbar icon items

Spend a little more time with Excel; it will look more and more different.

can be added or changed by clicking on it. Each command within Excel can have a toolbar button assigned to it.

Below the two toolbars, the next row offers a few different items that come in handy. The first area to the far left lists the cell address. That's pretty simple. But let's say the spreadsheet contains hundreds of cells full of information and you only normally refer to a few where some constant data resides. Let's say cells B34, Z66, and AA5 need constant attention.

In the cell address box, a user can change the name of any cell. Let's say cell B34 contains the interest rate on a short-term loan. Click inside the cell address box and type "Rate." Don't make it long; you can't use spaces or some special characters. Cell Z66 contains a recent salary bonus, so name it "Bonus," and AA5 holds how much you owe on the loan, so let's name that one "Loan." Your sizeable bonus means you might be able to pay off that loan, but you're not certain the extra interest will let you pay it off. Now, in an empty cell type

=Bonus-((Loan*Rate)+Loan)

and hit **Enter**. The formula may look a little like something from your old eighth-grade algebra textbook. It takes the loan amount, figures out the cost of interest, adds the interest to the loan principal, and deducts it from your bonus. If the number comes out whole, pay off the loan. Naming cells means no searching for any cells. Excel figures it out for you.

Just off to the right of the cell address is a button that looks something like a cursive fx, like this: *fx* That button accesses a new feature in Excel — the **Insert Function Search**.

> *The formula may look a little like something from your old eighth-grade algebra textbook.*

Aside from the standard addition, subtraction, multiplication, and division, Excel includes an impressive list of formulas and functions. From the summation or average of many numbers to depreciation of assets, from cosine or other trigonometry to logical functions with "If/Then" statements, you will find it here.

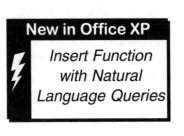

New in Office XP

Insert Function with Natural Language Queries

Excel includes an impressive list of formulas and functions.

The Insert Function command makes it easy. Create a new worksheet by hitting **File** then **New**, or you can just hit **CTRL+N**. Now, let's fill up all of the visible cells with random numbers. Hmmm…wonder if there's a function that might help us with that.

STEP 1: Click on the *fx* and simply type "How can I get a random number?" and hit the **Go** button and Excel will recommend the RAND function. But say you are not certain how RAND works. Click the little blue text in the bottom left corner that says **Help on this Function**. Excel launches a window so you can see that the RAND function generates a number greater than zero but less than 1 (*Figure 4.4*).

STEP 2: In order to create a screen full of numbers between 1 and 100, let's take RAND and multiply it by 100. Close all the helpful windows in your way and go to cell A1. Type in the following:

=RAND()*100

Hit **Enter**. A number like "81.19016417" should appear (of course, since it's randomized, your number will be different).

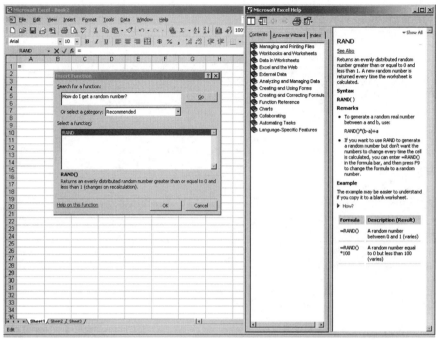

Figure 4.4

STEP 3: Now, click cell A1 and take hold on the heavy dot on the bottom right; left-click and hold, dragging it all the way to the left and let go. The first row should be filled with random numbers now.

STEP 4: Grab the same cell, hold and drag it all the way down leaving only one row empty. What you end up with is a screen full of numbers (*Figure 4.5*).

STEP 5: To make all these numbers more manageable, let's hide the numbers to the right of the decimal. With everything still highlighted, right-click and go to **Format Cells...** from the menu selection.

STEP 6: On the **Number** tab, select the word **Number** in the left box. To the right, a box will appear that says how many decimal places should be on screen. The default here is 2, but let's take that down to zero then click **OK**.

A box will appear that says how many decimal places should be on screen.

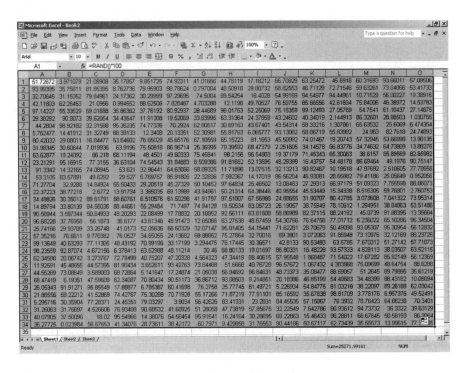

Figure 4.5

While you don't see the decimal places, the numbers are still there. Go to the first empty cell on the bottom row. Just out of curiosity, let's find out how many cells contain a number less than 50.

STEP 1: Click the *fx* and when the **Insert Function Search** box pops up, just ask it. Microsoft has built in "natural language" queries, meaning you can ask the question just as you would to a human being. So type: "How can I tell how many cells contain a number less than 50?"

STEP 2: This time, more than one choice shows up. Since we want to count the number of cells, the first few options are probably the most likely:

DCOUNT, COUNTA, COUNT, COUNTBLANK, and COUNTIF.

STEP 3: Highlight each one to see a brief description of the function. We don't want to count blank cells, so COUNTBLANK is ruled out. Also, the DCOUNT function only looks at entries in a single column, so drop that one. That leaves COUNTA, COUNT, and COUNTIF.

STEP 4: If you read about COUNTA and COUNT, both seek for specific entries. Only COUNTIF meets our requirement. Like before, click on **Help on this Function** to get the full details. Type into the cell "=COUNTIF(" then select the full range of cells. In my case, I will start at cell A1 and drag to cell O34. Don't hit Enter yet!

The DCOUNT function only looks at entries in a single column.

STEP 5: Type the following:

, (comma)
" (double quote marks)
< (less-than symbol)
50 (fifty)
" (double quote marks)
) (close parenthesis)

Now hit **Enter**. Your entry looks similar to the following (although your range of cells may vary):

=COUNTIF(A1:O34,"<50")

A whole number of some sort appears. Roughly half of the cells should be less than 50. Of course, we have no idea how many cells there are. So, in the next cell, let's find out. The plain COUNT function should help us with that. Type this in:

=COUNT(A1:O34)

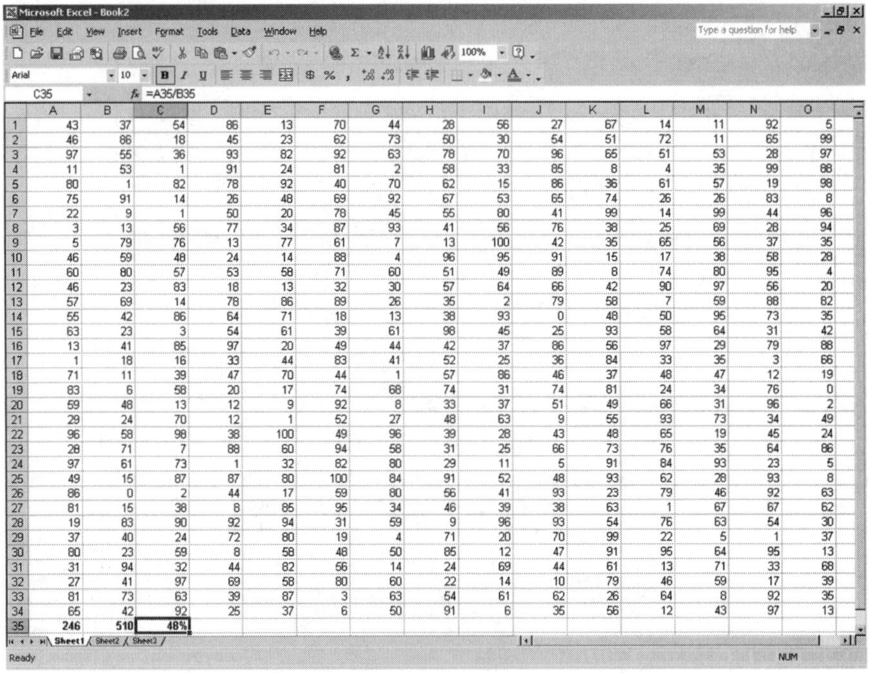

	A	B	C	D	E	F	G	H	I	J	K	L	M	N	O
1	43	37	54	86	13	70	44	28	56	27	67	14	11	92	5
2	46	86	18	45	23	62	73	50	30	54	51	72	11	65	99
3	97	55	36	93	82	92	63	78	70	96	65	51	53	28	97
4	11	53	1	91	24	81	2	58	33	85	8	4	35	99	88
5	80	1	82	78	92	40	70	62	15	86	36	61	57	19	98
6	75	91	14	26	48	69	92	67	53	65	74	26	26	83	8
7	22	9	1	50	20	78	45	55	80	41	99	14	99	44	96
8	3	13	56	77	34	87	93	41	56	76	38	25	69	28	94
9	5	79	76	13	77	61	7	13	100	42	35	65	56	37	35
10	46	59	48	24	14	88	4	96	95	91	15	17	38	58	28
11	60	80	57	53	58	71	60	51	49	89	8	74	80	95	4
12	46	23	83	18	13	32	30	57	64	66	42	90	97	56	20
13	57	69	14	78	86	89	26	35	2	79	58	7	59	88	82
14	55	42	86	64	71	18	13	38	93	0	48	50	95	73	35
15	63	23	3	54	61	39	61	98	45	25	93	58	64	31	42
16	13	41	85	97	20	49	44	42	37	86	56	97	29	79	88
17	1	18	16	33	44	83	41	52	25	36	84	33	35	3	66
18	71	11	39	47	70	44	1	57	86	46	37	48	47	12	19
19	83	6	58	20	17	74	68	74	31	74	81	24	34	76	0
20	59	48	13	12	9	92	8	33	37	51	49	66	31	96	2
21	29	24	70	12	1	52	27	48	63	9	55	93	73	34	49
22	96	58	98	38	100	49	96	39	28	43	48	65	19	45	24
23	28	71	7	88	60	94	58	31	25	66	73	76	35	64	86
24	97	61	73	1	32	82	80	29	11	5	91	84	93	23	5
25	49	15	87	87	80	100	84	91	52	48	93	62	28	93	8
26	86	0	2	44	17	59	80	56	41	93	23	79	46	92	63
27	81	15	38	8	85	95	34	46	39	38	63	1	67	67	62
28	19	83	90	92	94	31	59	9	96	93	54	76	63	54	30
29	37	40	24	72	80	19	4	71	20	70	99	22	5	1	37
30	80	23	59	8	58	48	50	85	12	47	91	95	64	95	13
31	31	94	32	44	82	56	14	24	69	44	61	13	71	33	68
32	27	41	97	69	58	80	60	22	14	10	79	46	59	17	39
33	81	73	63	39	87	3	63	54	61	62	26	64	8	92	35
34	65	42	92	25	37	6	50	91	6	35	56	12	43	97	13
35	**246**	**510**	**48%**												

Figure 4.6

The recalculation key causes Excel to generate all new random numbers.

After you hit **Enter**, you have your answer. Of course, it's not exactly half. Want to find out what percentage it is? In the next cell, type = (equals sign) then click on the first cell with the COUNTIF function, type **/** (slash) then click on the second cell with the COUNT function and hit **Enter**. Of course, the result appears to be zero, but it's not!

Highlight the cell and then click on the **%** (percent) symbol in the toolbar, making the number a percentage.

I ended up with 48 percent. Want to watch that number change? Hit the **F9** function key. That's the recalculation key and causes Excel to generate all new random numbers. Play with it a while and you'll notice that the results don't vary much, hovering between 46 and 55 percent, just as it should be with random numbers (*Figure 4.6*).

Just to the right of the *fx* you will see another blank area. Click any cell containing a random number and you'll notice that the contents beside the *fx* will change to the formula. This area shows the formula content of any cell. Edits can be made in this area without regard to formatting.

Beneath all the cells, three tabs appear named "Sheet1," "Sheet2," and "Sheet3." Change any tab by double-clicking it. Let's call "Sheet1" the "**Random Numbers**" sheet, then hit **Enter**. And these boring, plain tabs won't do, either. Right-click the tab and select **Tab Color** from the list. Make the tab any color

New in Office XP

Colored Sheet Tabs

you want. Let's go ahead and name "Sheet2" something like "**Making a Table**." Go ahead and give it a color, too.

Now, we've pretty much finished the basics of the spreadsheet interface, so it's time to build some information that's not necessarily numbers.

Click any cell containing a random number and you'll notice that the contents beside the fx will change to the formula.

Not Just Numbers

Most of us do not need to tap into the trigonometric functions, depreciation, or anything remotely statistical. But just because we've forgotten almost everything since taking the SATs doesn't mean a spreadsheet can't be extremely handy, even without numbers.

Here's an example for creating a table. In this case, it's a schedule.

Turn the text up on end to make the cells ready for thinning down.

You're captain of Spaceship Spreadsheet. For the next four weeks, your team of officers will be manning the spaceship's main bridge. Since you're the captain, you get to make the schedule. You've got six people on your crew: Yancy, Pete, Ruby, Garland, Betty, and Olivia. There are four shifts every day: midnight to 6 a.m., 6 a.m. to noon, noon to 6 p.m., and 6 p.m. to midnight.

Since the spaceship is in orbit this week, only one crew member has to be on the bridge, giving two people at a time the day off. So let's set up a basic grid.

In cell B1, type the word "Monday." Good news. Excel understands what you mean by "Monday," so it figures "Tuesday" must be next, and so on. Grab the holder and drag it out several cells to the right until you've got four weeks' worth — that happens at cell AC1.

Now here's a problem: Everything is so wide. How can we print out the schedule on a single sheet? All of the cells from B1 to AC1 should still be highlighted. Let's turn the text up on end to make the cells ready for thinning down. Right-click on the highlighted area and go to **For-mat Cells**. Choose the **Alignment** tab and go to the right-most window under the heading **Orientation**. Grab the red diamond by left-clicking and holding, sliding it up to the top, and then clicking the **OK** button (*Figure 4.7*). Text should be standing on end now.

But, we need to slim down the width. Click on the label row for columns B through AC (when you hover over this with your cursor, it turns to a solid down arrow). After selecting all those columns, choose one and make it smaller; from the right side, slide it left.

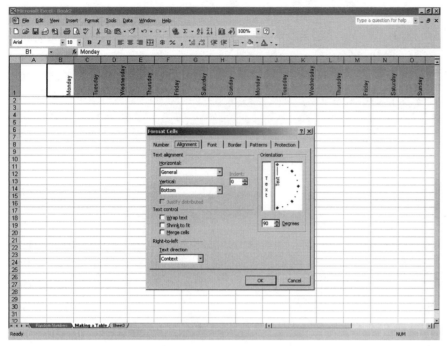

Figure 4.7

Notice that Excel lets you know how wide it is in its traditional measurement and a new pixel count. Narrow the cells to 20 pixels or 2.14. That should get all four weeks on a single screen so you can see the whole schedule.

New in Office XP

Size of columns and rows in pixels

In column A, starting with cell A2, type in the times of the shifts. You might need to widen column A to make it all fit. When you get done, you should have something like what you see here in *Figure 4.8*. Since all of the members of our crew have names that begin with different letters, we'll just use a letter to represent each one. Type them into the spreadsheet so the crew can read the schedule easily.

Excel lets you know how wide it is in its traditional measurement and a new pixel count.

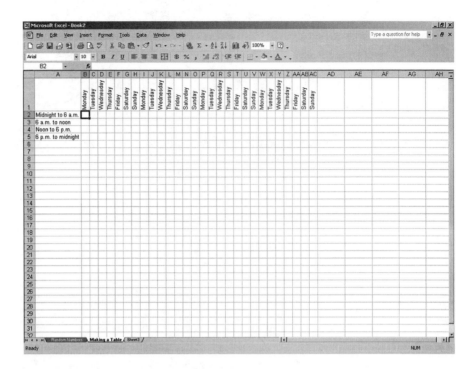

Figure 4.8

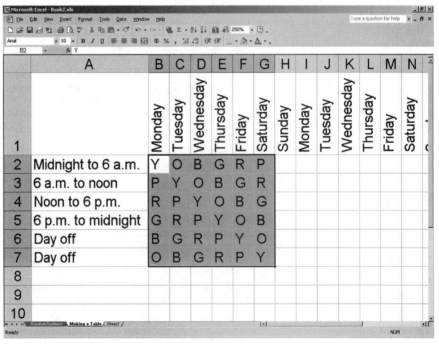

Figure 4.9

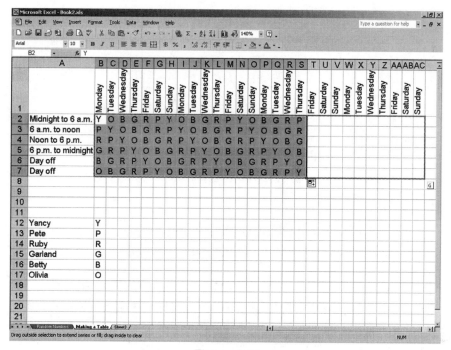

Figure 4.10

Let's go ahead and fill in the schedule. Type in a rotating schedule (*Figure 4.9*). Once the first six days are completed, the schedule repeats. No need to type it in. Just grab the bottom left corner and drag it out to the end of the schedule (*Figure 4.10*). Excel notices the pattern and simply repeats it.

But, let's admit it. This schedule is hard to follow.

To make it even easier, maybe we could assign colors to each of the crew members. The icon that looks like a paint bucket on the toolbar will come in handy for that. Just click the little down arrow next to the bucket and choose the right color whenever you've selected one of our crew's cells (*Figure 4.11*). Here's our guide:

Just click the little down arrow next to the bucket and choose the right color.

Yancy = **Y** = YELLOW
Pete = **P** = PURPLE
Ruby = **R** = RED
Garland = **G** = GREEN
Betty = **B** = BLUE
Olivia = **O** = ORANGE

Format the key so that each single-letter box has a color.

Format the key so that each single-letter box has a color (*Figure 4.11*). That formatting needs to be done for the next step.

Changing all the entries consumes too much time and would be tremendously tedious. So let's use a shortcut.

STEP 1: Pull up Find-and-Replace (**Edit** menu and click on **Find**. Or you can press **CTRL+F** to pull it up).

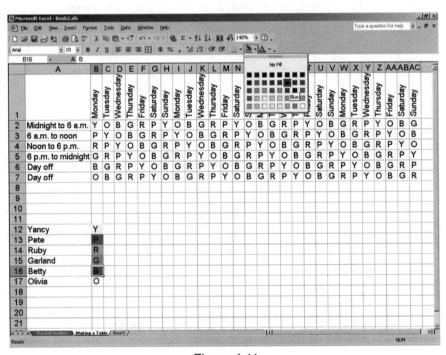

Figure 4.11

STEP 2: Click the **Replace** tab and then the **Options>>** button in the bottom left corner.

STEP 3: We're not going to change the contents of the cell, just the format. So, enter the letter of the person in both the Find and Replace blanks — in this case, let's do Ruby. So, put "R" inside both boxes.

STEP 4: Click the box **Match entire cell contents**. This prevents the Find-and-Replace function from noticing the "R" in Thursday, Friday, or Saturday.

STEP 5: To the far left of the Replace box (left of the area that says "Format Not Set"), click on the down arrow next to the **Format...** button.

STEP 6: When the menu drops down, click on **Choose Format From Cell...** (*Figure 4.12*).

Enter the letter of the person in both the Find and Replace blanks.

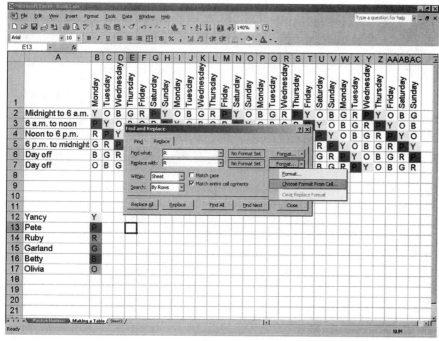

Figure 4.12

STEP 7: The Search-and-Replace box disappears for a moment. The cursor looks like a plus sign and an eyedropper together. Now, find and click on the already-colored R box in the key.

STEP 8: The Search-and-Replace box reappears. You'll notice the area that had said "Format Not Set" now holds "Preview*" inside a red box, just like the one you clicked on. Now click on the **Replace All** button.

STEP 9: You've done it. Excel notifies you how many replacements were made. Click **OK** and move on to the next member of staff until all are filled in.

Excel notifies you how many replacements were made.

All done. We have a table of the staff schedules, as seen in *Figure 4.13*.

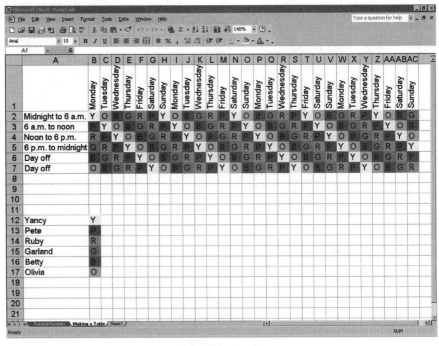

Figure 4.13

Formatting for a New World

In the above example, we did some formatting. Excel 2002 includes an ability to make tables — whether they contain text or numbers — easier to read with different formatting settings.

To demonstrate this by including one of the newest features, click on the third tab along the bottom that's probably named "Sheet3." Change the name to "Formatting" and set your tab color.

While on Spaceship Spreadsheet, some visitors need basic information on the location of important rooms. Let's create a table directory for them.

Excel XP includes an ability to make tables.

STEP 1: Starting in cell A2, type each of these entries on in separate rows and columns:

Location	Deck	Section	Rm
Mess Hall	Deck 3	Aft	10
Command Center	Deck 1	Forward	11
Recreation Center	Deck 4	Starboard	5
Sick Bay	Deck 2	Aft	2
Visitor's Quarters	Deck 5	Port	3

STEP 2: You've probably got a table that looks something like *Figure 4.14*. Of course, the text appears cut off since the column isn't wide enough. Highlight the entire spreadsheet by clicking the small box in the upper left corner between the headings for column A and row 1. Hover the cursor close to the right border of column A where the cursor turns into a symbol like this ←|→ (two arrows pointing in opposite directions with a line down the middle). Double-click there. This causes an automatic adjustment to all the columns based

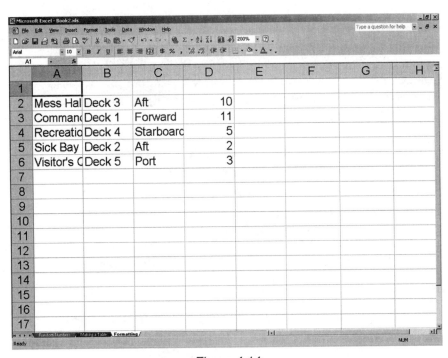

Figure 4.14

Adjustments can be made to individual columns or rows, based on what is highlighted.

on the widest contents (see *Figure 4.15*). Such adjustments can be made to individual columns or rows, based on what is highlighted.

STEP 3: While everything is highlighted, let's change the font. The toolbar with fonts should be up. Just change the font to Tahoma or something else. Everything in the spreadsheet is changed.

STEP 4: Highlight row 1; let's make the font slightly smaller in that row—say 8 points. Let's also bold that line by hitting the **B** button in the **Formatting** toolbar (*Figure 4.15*).

STEP 5: Highlight cells A1 to D1 — just the four cells that contain the head entries of the directory. Click on the arrow next to the paint bucket — known as **Fill Color**. A palate of colors drops

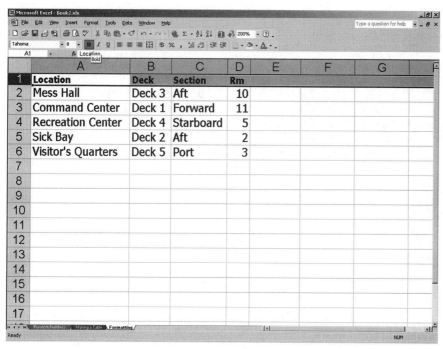

Figure 4.15

down. You can hover over each one a moment and a little box appears with the color. Select black in the top left corner.

STEP 6: Since the font was black, we normally couldn't read the contents of the cells, although since they're highlighted, we can read them somewhat. So let's change the font color with the button next to the paint bucket—it looks like an **A** with a bold, blue underline. Click the down arrow next to it and another palate drops down. Now click the light yellow sample (*Figure 4.16*).

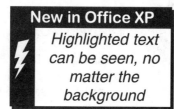

New in Office XP

Highlighted text can be seen, no matter the background

You can hover over each one a moment and a little box appears with the color.

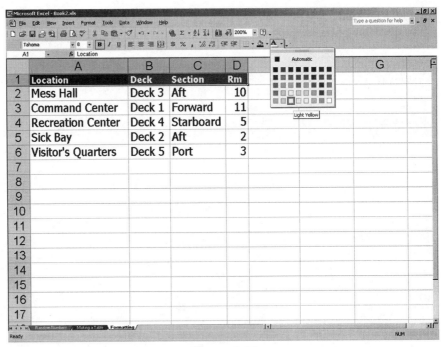

Figure 4.16

The first button that looks like a pencil allows drawing of individual lines or an entire grid pattern.

STEP 7: Let's draw some borders around the rooms. In previous versions of Office, this process could be a little tedious. In Office 10, you can select a new border drawing tool. It's a toolbar. In the toolbar area, right-click, and in the menu select **Borders**. The first button that looks like a pencil allows drawing of individual lines or an entire grid pattern. Draw some lines around entries. Change line widths, patterns, and colors with the tool. Have some fun! (*Figure 4.17*).

New in Office XP

Border Drawing tool

STEP 8: Now, let's add some background color. Choose any single cell and hold down the **CTRL** key.

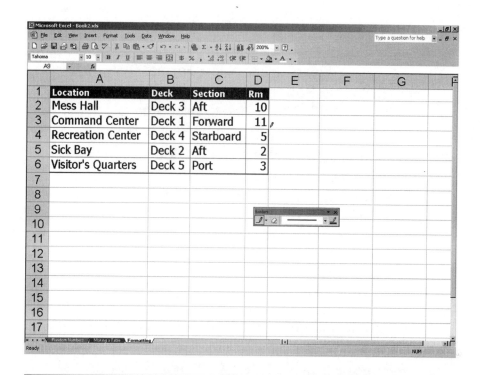

Figure 4.17

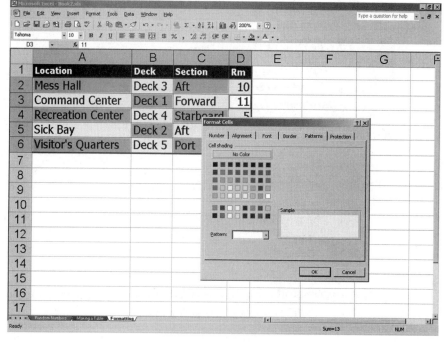

Figure 4.18

Microsoft provides a great deal of information built into Excel so you can tap into the power.

Select any other cell; it doesn't matter whether that cell is next to the other or several cells apart. Right-click and select **Format Cells...**. Click on the **Patterns** tab and choose a color (*Figure 4.18*). The feature to select noncontiguous cells for formatting premiered in Excel 98.

From the **Format Cells...** dialogue, you can see all of the formatting options available in Excel. There's plenty. Play around with it to discover just how fanciful your spreadsheets can get.

It All Started with Math

At its root, Excel remains a very powerful mathematical and statistical application. Earlier in the chapter, we looked at the new *fx* Insert Function Search tool. Excel offers 238 different functions, not to mention all the other mathematical operations with the **+**, **-**, *****, and **/** keys.

New in Office XP

Function Arguments provides "blanks" for every part of a function

The Insert Function Search also provides some step-by-step help. After finding a function to use, click the **OK** button and the **Function Arguments** dialogue box appears. The box lists all of the fields that need to be completed.

Don't worry if you get it wrong. A Smart Tag appears with an alert along with help menus.

A whole book could be dedicated to each function. Luckily, Microsoft provides a great deal of information built into Excel 2002 so you can tap into the power.

But, clicking the *fx* shouldn't need to be an option of first resort. We'll examine some of the more frequently used functions and other math basics. To get started, let's add another worksheet to the current spreadsheet.

STEP 1: Hover the cursor over the current colored tabs and right-click. When the menu pops up, select **Insert**....

STEP 2: Choose **Worksheet** from the items listed there and click **OK**.

STEP 3: As a head start, add several numbers to every cell from A1 to A20, just like in *Figure 4.19*. Make some big and others small, some whole and some with decimals.

You can choose multiple ranges or even single cells without any limit.

With that done, let's look at a short list of the most-often used — and some not-so-often but powerful — functions in Excel. *Figure 4.19* contains every function covered here. While some examples appear here, your entries may vary based on what cells you're performing functions on.

=SUM(A1:A20) simply adds up the contents of a range of cells. You can choose multiple ranges or even single cells without any limit. Let's say we only want cells A1 through A5, A7 through A10, A15 through A18, and cell A20. The formula would look like this: **=SUM(A1:A5,A7:A10,A15:A18,A20).** If you're typing it by hand, put a comma between the ranges or cells. If you want to select the cells with the mouse, hold the **CTRL** key.

=AVERAGE(A1:A20) finds the average, as if you'd added up all 20 cells and then divided by 20. Like the =SUM function, groups of noncontiguous cells can be selected.

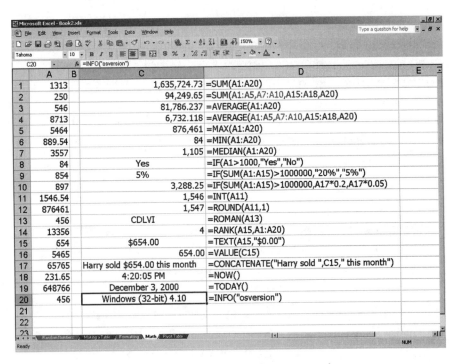

Figure 4.19

=MAX(A1:A20) returns the maximum number of a range, meaning the number with the most value (in *Figure 4.19*, you'll see it's 876,461). This allows, for example, a quick determination of which salesman sold the most last quarter. And like =AVERAGE and =SUM, groups of noncontiguous cells can be selected.

=MIN(A1:A20) would be the opposite of =MAX, returning the cell with the least value.

=MEDIAN(A1:A20) gives the mean or number in the middle of a range. If you don't remember mean from school, don't worry. Think of it as the awkward middle child. So in the range of numbers 1, 2, 3, 7, and 20, the median number would be 3.

=IF(A1>1000,"Yes","No") is a simple example of the If…Then…Else function available in Excel. In this formula, it asks if the value in cell A1 is greater than 1,000, then the word "Yes" appears. If it's less than 1,000, the word "No" appears. The =IF function can be very powerful. Let's say a salesperson has a goal of more than $1 million a year to get a 20 percent bonus of a certain salary. Otherwise, the salesperson gets a 5 percent bonus. Let's say cell A17 in *Figure 4.19* contains the salesperson's salary, and cells A1 through A15 contain the sales for that year. For informational purposes, here's the formula that returns what the bonus would be: **=IF(SUM(A1:A15)>1000000,"20%","5%").** So, if the sum of cells A1 through A15 is greater than 1 million, then the text "20%" appears. Otherwise, a "5%" shows up. To return the actual amount of the bonus, use this formula:**=IF(SUM(A1:A15)>1000000,A17*0.2,A17*0.05).** Same basic principle: if the sum of cells A1 through A15 is greater than 1 million, then multiply cell A17 by .2 (or 20%). If not, then multiply cell A17 by .05 (or 5%). The =IF function can be very powerful.

=INT(A11) rounds a number down to the nearest integer. In *Figure 4.19*, the cell A11 contains the number 1546.54. The =INT rounds that down to the whole number 1,546.

=ROUND(A11,1) rounds to the nearest whole number. In the case of 1546.54, the result is 1,547. Excel also contains functions like =ROUNDUP and =ROUNDDOWN.

=ROMAN(A13) will return text, turning the Arabic number in cell A13 to the Roman numeral. In the case of cell A13, 456 becomes CDLVI.

The =IF function can be very powerful.

=RANK can help when, in the case of a sales team, who sold the most and where each ranks as compared to the others.

=RANK(A14,A1:A20) returns the ranking of the number in a range. In other words, the value of cell A14 is 13,356, the fourth-largest number of the 20 numbers listed. Therefore, the returned value is 4. =RANK can help when, in the case of a sales team, who sold the most and where each ranks as compared to the others.

=TEXT(A15,"$0.00") will turn a number into text and format it as requested. In this case, the value in A15 was formatted as money with two decimal places. Choose your own based on the entry in the Number tab of the **Format Cells...** command.

=VALUE(C15) goes the opposite direction. It turns text into a number.

=CONCATENATE("Harry sold ",C15," this month.") will string together whatever text you'd like. In the example, the result is a sentence that looks like this: "Harry sold $654.00 this month."

=NOW() gives the exact time while **=TODAY()** returns today's date. Format the cell to present the date in any preferred style.

=INFO("osversion") returns the specified information. In this case, the requested value is the version of the operating system in use (Windows 98 is the same as Windows 4.10). Other options are "directory" for the path of the current folder, "numfile" for the number of active worksheets in the open workbooks, "release" for the version of Excel currently in use, and "totmemory" for the total amount of memory.

Absolutely Relative and Other Considerations

Cell selection for formulas can vary, as you may now realize. Excel calls the differences "absolute" at one end and "relative" at the other.

Take a look at *Figure 4.20*. The spreadsheet shows the results of sales from the last year for a sales team. In cell B6, click the toolbar button with the Greek Sigma character (Σ). This button is **AutoSum**, and it figures out that the column needs to be added up. When Σ was pressed, Excel showed **=SUM(B2:B5)** for confirmation. Just press **Enter** because Excel got it right.

Excel calls the differences "absolute" at one end and "relative" at the other.

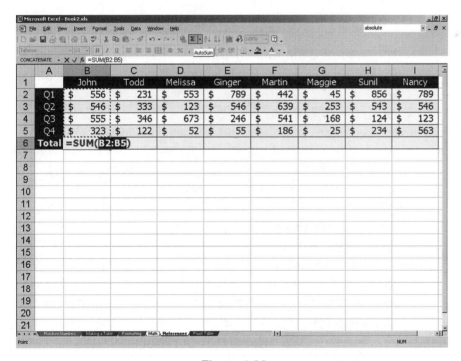

Figure 4.20

We can go through each column and press that Σ, but sometimes that would take a long time. Let's save time by using the **AutoFill** function. Highlight cell B6, go to that bottom left block, left-click and hold. Now, drag to the right until all the cells are highlighted.

Notice that Excel does not just add up cells B2 to B5. It moves the formula in a relative way, so in cell C6 is the sum of C2 through C5, cell D6 contains the sum of D2 through D5 and so on.

Save time by using the AutoFill function.

These are relative references. Excel handles them rather well. But sometimes an absolute is needed. In this case, let's say that each salesperson will receive a bonus based on sales. Put that bonus percentage in cell A7. Let's be generous and make it 20 percent.

This cell will remain constant — an absolute. It's simple enough in cell B7 to enter the formula **=A7*B6** to get the bonus amount. However, if we drag it over using AutoFill, cell C7 will contain =B7*C6, which gives a wrong result by multiplying John's bonus by Todd's sales.

Let's go back to cell B7. Type = and then click on cell A7, which contains the **20%**. Now, press the **F4** key. Did you notice how **=A7** becomes **=$A$7**? Go ahead and finish the formula with *, and click on cell B7 and hit **Enter**. Now, AutoFill the rest of row 7 (*Figure 4.21*).

Click on any cell in row 7. Notice each one still has **=$A$7** at the beginning. Hitting the F4 key made that reference absolute. No matter what, the cells will look to cell A7 for the formula.

Now, let's total up the bonuses (remember the Σ) we're giving and … oops! It's more than $2,000! We didn't

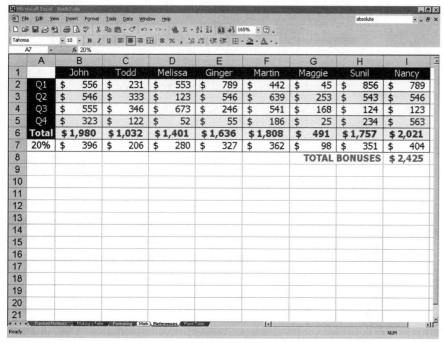

Figure 4.21

want to give out more than $2,000 this year. So, thankfully, because we used an absolute, just go in and change the number in cell A7. Dropping the bonus to **16%** puts the total cost at less than $2,000.

Powerful (but Confusing) PivotTables and PivotCharts

Among the most powerful analysis tools in Excel are the PivotTables and PivotCharts. The previous version of Excel contained PivotTables that could be made into charts, but not directly into PivotCharts in the same dynamic way as a PivotTable.

Among the most powerful analysis tools in Excel are the PivotTables and PivotCharts.

But, how in the universe can you figure out a PivotChart if you have no idea what a PivotTable is?

Honestly, as we can attest, we're at a loss as to how to describe anything "pivot" in Excel. We think it's better just to show you.

Figure 4.22 shows just a portion of the spreadsheet Linda made. Linda has been keeping track of all office supplies purchased by a small company over the last year. Notice she's got the employee's name, the employee's department, what month it was purchased in, what was purchased, from which store, how many, the unit price, and the total cost.

How in the universe can you figure out a PivotChart if you have no idea what a PivotTable is?

New in Office XP

Pivot Charts

So, Linda is wondering which department spent the most money. It's a big spreadsheet with more than 100 rows. She could sort the department row and then add up the totals. But then she might want to know which employee spent the most, which means more sorting and adding. Oh, and she's wondering to which store did most of the orders go. Again, this is a lot of work.

Or very little. PivotTables to the rescue!

Under the **Data** file menu, select **PivotTable and PivotChart Report....** A wizard pops up to step a user through the process. We'll start with a PivotTable first. With step two, the wizard automatically selects all the data in the columns.

PivotTables and PivotCharts only use data under a header column. It doesn't recognize data organized by

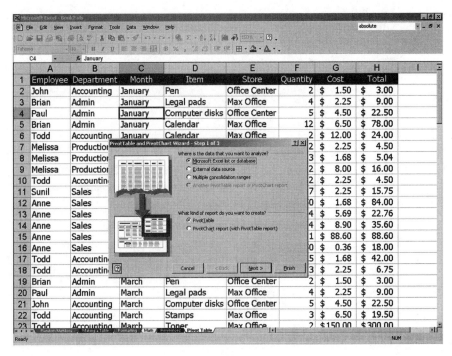

Figure 4.22

row. Keep this in mind. And make sure you have a header column — that's a column with the name of the contents in it, in this case Employee, Department, Month, Item, Store, Quantity, Cost, and Total. You'll understand why in just a moment. Step three finishes the Wizard, creating another worksheet by default.

Figure 4.23 shows the resulting table with the two floating palates. Do you see the palate named **PivotTable Field List**? It's got all those column headers.

Linda wanted to know which department spent the most. That's easy. From the **PivotTable Field List**, drag the **Department** box into the area with the text "Drop Row Fields Here." Then drag the **Total** box into the "Drop Data Items Here" area. *Figure 4.24* shows the result that

Make sure you have a header column.

117

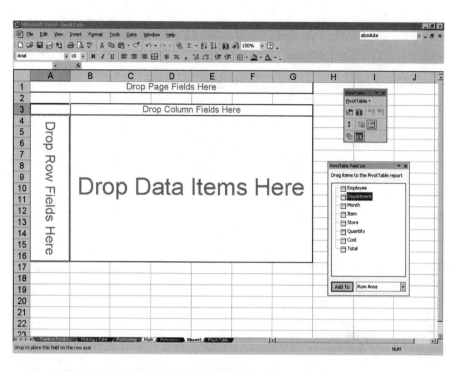

Figure 4.23

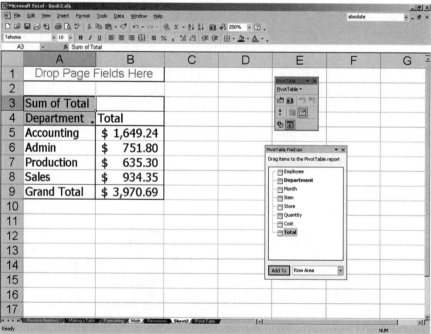

Figure 4.24

accounting spent the most money on supplies. Drag the solid **Department** box back into the **PivotTable Field List** and put the **Employee** box into place to see that Todd spent the most money.

Drag and drop numerous fields. See that more folders than anything else was purchased and the Max Office store had the most purchases. PivotTables provide a very powerful way to manipulate data quickly.

And so does PivotCharts. Go back to the original worksheet to create one. PivotCharts work the same basic way, and fields can be dragged into different areas to create a chart. See *Figure 4.25* for an example of what items were purchases from which store with the total amount spent shown too.

PivotTables provide a very powerful way to manipulate data quickly.

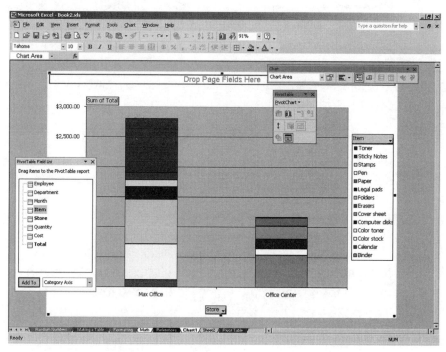

Figure 4.25

Charting a Course

To illustrate the charting function in Excel, let's grab some numbers from the original random number sheet. Highlight a bunch of numbers in the worksheet and hit **CTRL+C** to copy those numbers to the clipboard. Now create a new worksheet.

We just want the numbers, not the constantly changing random numbers.

We just want the numbers, not the constantly changing random numbers. So go to the **Edit** menu item and select **Paste Special….** A dialogue box pops up with several choices. Let's try the **Values** radio button. Now hit **OK**. That gives us the numbers with several decimal places. Hit **CTRL+Z** to undo. Now select **Paste Special…** again and choose the radio button **Values and Number Formats**. Before clicking OK, also check the **Transpose** box. The Transpose command can be very handy since it turns all columns into rows and all rows into columns.

New in Office XP

More **Paste Special…** options

Add some column headers, like MTWTF for the days of the week. Let's give each row a state name (*Figure 4.26*). Highlight a small selection of cells — let's say the first few states and the first week — and then click on the **Chart Wizard** button in the toolbar.

Make a bar chart, taking all the defaults, resulting in a chart something like *Figure 4.27*. Notice the colored boxes around the areas in the worksheet. Move that box around. The boxes can be made larger or smaller, as appropriate.

Notice how the F for Friday appears at the top. Double-click along the axis near the days of the week. When the

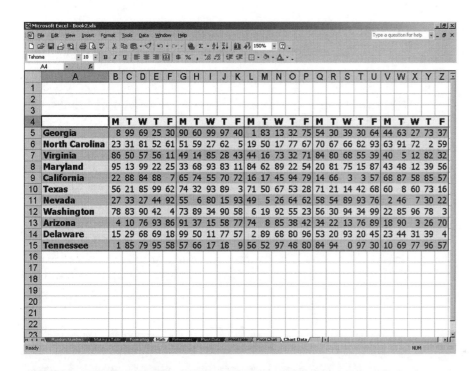

Figure 4.26

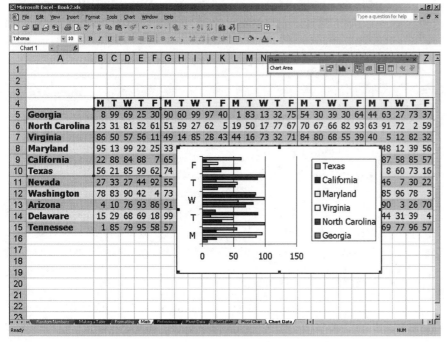

Figure 4.27

Format Axis box appears, select the **Scale** tab and click the box next to **Categories in reverse order** and click **OK**.

Also notice in the example that all the bars approach 100 but never go beyond. Yet the Chart Wizard chose to go to 150. Double-click that axis and select the **Scale** tab in the **Format Axis** box. Change the **Maximum:** entry to 100 and, also, let's put a few extra intermediate lines in by changing the **Major units:** entry to 25. Now click **OK** and the chart is starting to shape up.

Microsoft chooses a palate of colors that it finds appealing. But the muted defaults may not be perfect and do look very close to one another. While in the chart, click on any bar to select the whole series. Change the

Microsoft chooses a palate of colors that it finds appealing.

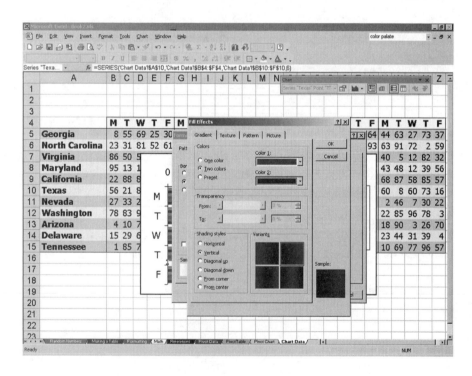

Figure 4.28

colors to make the hues be a little more varied so they are easier to distinguish.

Now let's look at the top winner — the one bar longer than all the others. Let's say that bar needs to be highlighted. Click the bar once and the whole data series is selected. Pause a second. Click the same bar again. Now, just the bar itself should be highlighted. Double-click it to get the **Format Data Point** dialogue box. Now, that entry can look a little different, perhaps with a **Fill Effect** (*Figure 4.28*).

Charting in Excel is a powerful tool. Don't hesitate to try it out. And like all of the other applications in Office, charting uses the same formatting process.

Click the bar once and the whole data series is selected.

5 Outlook

Communicate with the far reaches of the cosmos or just down the hallway. If a coworker, relative, or friend has an e-mail address, Microsoft Outlook 2002 can keep in touch with him or her — and with your life. Whether it's scheduling the next shuttle launch or your daughter's soccer practice, Outlook can keep you on top of life's challenges.

Even though Outlook's life as a member of the Office suite has been relatively short, some companies and individual professionals find the information manager a requirement. Outlook's primary tasks are e-mail, scheduling, and contact management. Outlook offers some other features that we will cover briefly, but the focus is mainly on those three.

We have been Outlook users for a few years now, as Outlook first appeared as part of Office 97. After working out some quirks, Microsoft released a special version of Outlook 98 even though there was no Office 98 (except for the Macintosh platform). When Outlook 98 showed up, we were hooked forever. Outlook 2002 continues to build on all previous versions of the application.

Outlook offers many little nuances that conform to the way you work.

Basic Concepts

Outlook is not hard to understand. Just take any paper address book, add a calendar, and throw in e-mail. Okay, so maybe you can also add those little yellow sticky notes and a to-do list as well. Oh, we also need a box of crayons and some maps. And a diary — we should add that too. Oh what the heck, throw in a personal assistant!

Obviously, Outlook offers many little nuances that conform to the way you work. It also may force you to work more efficiently.

Once upon a time, applications like Outlook were called "personal information managers." Not so anymore. Outlook can manage more than personal information. A business team, a whole company, or a single parent — virtually any group or individual can use Outlook to manage life.

A Constantly Changing Interface

When you open up Outlook 2002, you are presented with something that resembles *Figure 5.1*. The **Outlook Today** screen presents the "at-a-glance" of your schedule for the coming week, tasks to be achieved, and the inbox of new e-mail messages.

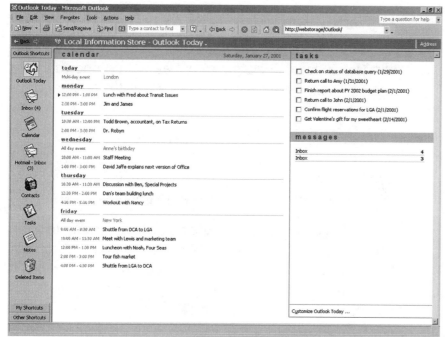

Figure 5.1

Outlook looks and feels like any basic schedule / address book in the physical world.

Outlook Shortcuts appear along the left side and the menu items **Eile**, **Edit**, **View**, **Favorites**, **Tools**, **Actions,** and **Help** appear along the top.

Outlook's interface changes based on which part of the application is being used at the time. Click on any of the icons in **Shortcuts** and the interface changes accordingly.

Outlook looks and feels like any basic schedule/address book in the physical world.

Compared to the other applications in Office, Outlook got only minor work. We'll cover all that's new, but with this chapter, we thought we'd also focus on the features we get a lot of questions about.

E-mail, E-mail, Everywhere!

Many people may think that e-mail is the primary reason for Outlook's existence, but as will be shown in the coming sections, e-mail integrates well with the other components.

Outlook supports most e-mail types.

How to set up an e-mail account in Outlook varies based on the type of account. Outlook supports most types, whether you are getting e-mail at home from an Internet service provider or at work from the local server.

In the **Outlook Shortcuts** bar, click on **Inbox** to open your mailbox. You might have more than one Inbox, like the two shown in *Figure 5.2*. Two mailboxes are helpful if you want to separate business and personal e-mail accounts. Let's get an account set up first.

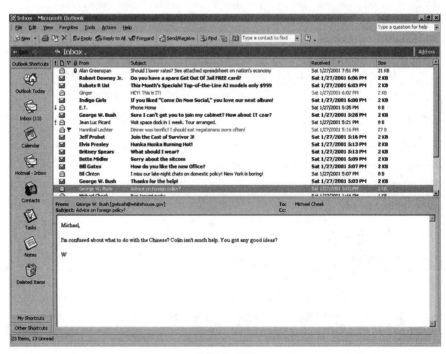

Figure 5.2

Exchange with the network

In the workplace, Outlook can be the standard application for e-mail. Outlook is the client side of an enterprise application by Microsoft known as Exchange Server. Exchange shuttles e-mail, appointments, and more among members of an organization. Exchange can also connect to the Internet and act as a kind of gateway to the world.

A network administrator who is specially trained should perform the account setup process in Exchange.

Network installations of Outlook store all data within Exchange's massive database on a network server. Other than the application and a few data files, most is not stored on your personal computer.

Exchange shuttles e-mail, appointments, and more among members of an organization.

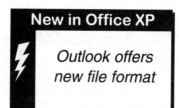

New in Office XP

Outlook offers new file format

You may want to consider creating a Local Information Store — .LIS file; previous versions of Outlook used .PST files or Personal Folders File. Outlook 2002 can use both of these file formats. If you want to store some mail files on your PC, perhaps as a backup or to stop the annoying warning that your mailbox is too big, consider creating an .LIS file.

From the **Tools** menu, select **Data File Management**.... Click the **Add...** button and choose either .LIS or .PST. Both are compatible with Office XP.

Internet

In a stand-alone mode without Exchange on the back end, Outlook works just as well. Outlook supports most standard e-mail types — the most common being POP3 and SMTP. These acronyms stand for Post Office Protocol 3.0 and Simple Mail Transfer Protocol. If you have an Internet service provider, you probably use POP3 to receive e-mail and SMTP to send it.

If you have an Internet service provider, you probably use POP3 to receive e-mail and SMTP to send it.

New in Office XP

Outlook supports Hotmail

Outlook now supports Hotmail, a free Internet e-mail service from MSN, the Microsoft Network. To create an e-mail account, go to the **Tools** menu and select **E-mail Accounts...**. Outlook also offers a new wizard for creating accounts. The process is easier than before.

Select **Add a new e-mail account** and click the **Next >** button. On the next screen, select the type of account. As mentioned before, **Microsoft Exchange Server** is first on the list. **POP3** is next, standard for most Internet service providers. **IMAP** stands for Internet Message Access Proto-

New in Office XP

Wizard for creating new e-mail accounts

col, which is similar to POP3 but used less often. **HTTP** is next, which stands for HyperText Transfer Protocol. You might recognize this from a Web browser where **http://** always appears before an address. This is the type used by Hotmail. Finally, the Additional Server Types are listed, but choosing one of those generally requires someone with specific knowledge of your network.

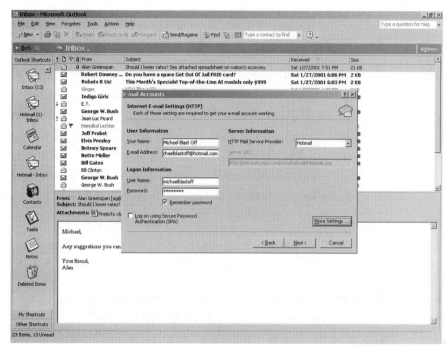

Figure 5.3

A window may pop up asking you to verify the password.

Click on **HTTP** and then click the **Next >** button. As in *Figure 5.3*, choose a name, e-mail address, user name, and password. In this case, **Michael Blast Off** is the name, **michaelblastoff@hotmail.com** is the e-mail address, and the user name is **michaelblastoff.** As should be the case, the password is a secret! A window may pop up asking you to verify the password. The **Remember Password** box can be clicked so Outlook keeps the password in memory. If you're the only one who uses the computer, this is very convenient. If not, you will probably want to leave this unchecked. Click **OK**.

Outlook downloads the appropriate folders — like inbox, outbox, sent, etc. — and opens a windowpane just right of the **Outlook Shortcuts** bar. This windowpane, known as the **Folder List**, shows all folders Outlook currently uses.

From Local Information Stores and Personal Folder Files to Hotmail accounts, everything is here.

In our example, the new account showed up as "Hotmail (1)", which is not very descriptive. By right clicking on **Hotmail (1)** and selecting **Properties for "Hotmail (1)"...**, you can make changes. To change the name, click on the **Advanced...** button and change the text in the **Name:** field. Click **OK** and **OK** again and it's changed.

Outlook can also recognize e-mail addresses as each is typed into these fields.

Creating e-mail

Microsoft has added a couple more minor enhancements to e-mail. When creating a new e-mail message, begin to type in the recipient names in the TO:, CC:, or BCC: fields. You'll notice as you type, Outlook guesses what name you're entering. For example, a contact named Sabrina might be in your contact list. By the time you type in the "b," Outlook will probably list Subrina in a small box below the field. Simply hit EN-TER to accept that choice. Outlook automatically inserts the e-mail address. Outlook can also recognize e-mail addresses as each is typed into these fields.

New in Office XP

Auto-completes names and e-mail addresses

If this feature becomes bothersome, it can be turned off. Under the **Tools** menu, select **Options**... and click the **E-mail Options**... button on the **Preferences** tab. Finally, click the **Advanced E-mail Options**... button. Several different settings are listed here, but you'll find **Suggest names while completing the To, Cc, and Bcc**

fields in the bottom group of items. Turn this item by clicking to put a check in the box or turn off the item by unchecking the box the same way.

Other settings within these dialogues might be helpful. For example, some people enjoy using fonts that are not always easy to read, such as something cursive or annoying. You can put a stop to reading e-mail containing these difficult-to-read fonts!

Under the **Tools** menu, select **Options...** and then select the **Mail Format** tab. Click the **Fonts...** button about halfway down. Here you can choose which fonts to use. The bottom section allows you to select **Always use my fonts**, which means that your incoming and outgoing mail will always use the font choices you make.

Outlook now defaults to an HTML format for all e-mails. HTML contains the hidden codes that web browsers use. Because Outlook uses Word as its e-mail editor, as you're typing, Word can check spelling and grammar on the fly if the feature is enabled. You'll also notice some of the same toolbars along the top of Outlook as are contained in Word.

HTML is a powerful format and Word makes it easy. But sometimes e-mail cannot be sent in that format or a recipient cannot see HTML-encoded messages. For that reason, you may need to convert messages to rich-text format, also known as RTF, or even plain-text format. In the Outlook toolbar, which begins with the **Send** button and appears beneath the Word toolbar(s), the last item is a drop-down menu that probably says

You can put a stop to reading e-mail containing difficult-to-read fonts!

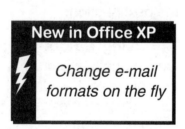

New in Office XP

Change e-mail formats on the fly

HTML▼. Just click the down arrow to switch to RTF or plain-text format.

Such a change can be made permanently. Return to the **Tools** menu, select **Options...**, and then select the **Mail Format** tab. Another drop-down menu at the top of this dialogue allows the choice of whether to compose in HTML, RTF, or plain-text format. If your PC is a little too slow or lacks memory, you can also turn off the setting for **Use Microsoft Word to edit e-mail Messages**.

Once the e-mail starts to flow in volume, managing all of it can become a challenge.

Organizing e-mail

Once the e-mail starts to flow in volume, managing all of it can become a challenge. Outlook offers some basic organizational tools to make it easier. In the toolbar to the right of the **Find** button is an icon that looks like several yellow blocks with the label **Organize**. Click it.

A new windowpane appears with four options along the left, as seen in *Figure 5.4*. With the **Using Folders** function, messages can be moved to different folders automatically. This creates rules, which are discussed in the next section.

Using Colors helps highlight certain messages. Outlook uses the current highlighted message to create the color rule. It is all relatively straightforward to create a rule so that all messages from a particular person appear in the color teal or even purple.

Using Views allows you to sort messages by a variety of ways. Just double-click the options to see what each looks like and then leave it set on the one that you like the best.

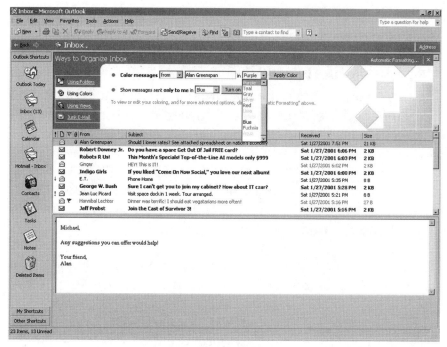

Figure 5.4

Finally, the handy **Junk E-mail** option will color-code or delete spam messages or those that contain adult content. We prefer to delete these messages. Change the drop-down menu that currently says **color** to **move**. Make sure the other drop-down item in the sentence says **Deleted Items**. The phrase reads, "Automatically move Junk messages to Deleted Items." Click the **Turn On** button to activate this option. You might want to do the same for the Adult Content messages.

Click on the "For more options, click here." You can learn more about how to deal with junk e-mail and adult content. Outlook can automatically delete any such messages per your designation. For example, say that some spam-generating company keeps sending you junk e-mail and you cannot seem to stop it. Right-click on the

Outlook can automatically delete junk e-mail messages per your designation.

135

message. From the drop-down menu, select **Junk E-mail ▼** and then choose **Add to Junk Senders list**. From now on, a message from that e-mail address is shuttled off to the **Deleted Items** folder. You can check that folder occasionally to make sure you're not deleting something you needed. **Deleted Items** is flushed occasionally.

If you have not accessed this feature before, you might get a warning the first time you try to make changes. To stop this warning message from appearing in the future, click the box next to **Please do not show me this dialogue again** and then click **OK**.

The flag for follow-up dialogue allows the creation of an item that will remind you to accomplish a task.

Have you gotten a message from someone that you need to remember to write back or perform some other follow-up action related to the message? Follow-up flags can remind you. Right-click on any message and select **Follow Up...** from the drop-down menu. The flag for follow-up dialogue allows the creation of an item that will remind you to accomplish a task, return an e-mail, or even for informational purposes only, so that you can quickly access an e-mail requiring attention at an appointed time. A reminder will chime and a small screen will pop on the display when the appointed time arrives.

Rules of engagement

We receive hundreds of messages a day via e-mail. Although the organize tools in Outlook provide a beginning point, so many messages may require a more comprehensive set of rules.

Outlook provides rules to perform almost any basic task. Let's begin by creating an extra folder with some rules to make e-mail easier to manage.

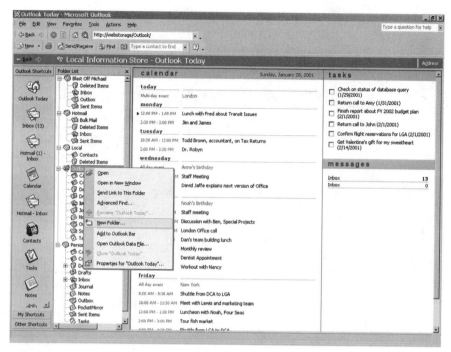

Figure 5.5

STEP 1: Right-click on **Outlook Today** or a **Local Information Store** in the **Folder List** (*Figure 5.5*). Select **New Folder...** from the drop-down menu.

STEP 2: A window like the one shown in *Figure 5.6* pops up. Select the **Mail and Post** item. In the **Name:** field, type in your name. In this example, we're calling it **Finance**. Click **OK**. A prompt will ask whether you want to add the folder to the **My Shortcuts** area of the **Outlook Bar**.

STEP 3: The idea is for all Finance-related e-mail to move to that folder automatically. For example, any e-mail from accountant Todd Brown, CPA, would go into that folder. To set this up, under the **Tools** menu, select **Rules Wizard....**

All e-mail related to a certain subject can be moved to a designated folder automatically.

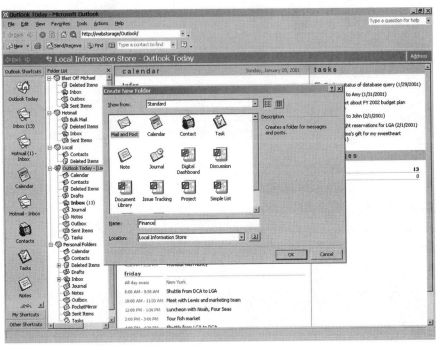

Figure 5.6

You might get a warning about rules from HTTP mail not able to be processed.

You might get a warning about rules from HTTP mail not able to be processed. Click the **OK** button if you see it. Essentially this means that if Mr. Brown sent e-mail to your Hotmail account, the rules wouldn't work. From other accounts like those using Exchange and POP3 mail, this feature works fine.

STEP 4: The **Rules Wizard** should be up for your primary e-mail account, which is listed in the top field next to **Apply changes to this folder:**. Click on the **New...** button.

STEP 5: Rules can be created in two ways — from

a template or from scratch. Templates are common rules, like the simple one we're now working on. As a matter of fact, notice the current template rule highlighted known as **Move new messages from someone**, which is exactly what we're doing. In the bottom field, see the rule as it stands:

Apply this rule after message arrives from <u>people or distribution list</u> move it to the <u>specified</u> folder.

STEP 6: Click on **<u>people or distribution list</u>**. Since Todd Brown CPA is listed in my contacts, I just need to select him from the list. Then click **OK**.

STEP 7: Click on **<u>specified</u>**. Find the Finance folder in the Local Information Store, select it, and then click **OK**.

STEP 8: Now the rule reads,

Apply this rule after message arrives from <u>Todd Brown CPA (tbrown@biacpa.net)</u> move it to the <u>Finance</u> folder.

Now click the **Finish** button.

Rules can be much more powerful than this basic example, and even more granular. Let's say you're working on a project code-named SuperSpaceShip. You can set up a rule that takes any messages with the word "SuperSpaceShip" and deposits them into a Project folder. The rule can also send a copy to your boss, except for messages from your boss. Messages from your boss can be highlighted as a priority and the computer set to play a tune whenever one of these priority messages arrives.

Messages from your boss can be highlighted as a priority.

This is a rule created from scratch, or as the Rules Wizard calls it, a "Blank Rule." Rules are relatively easy to create. Just go through the steps one at a time and select the underlined text to customize the rule.

Signature files

The signature usually appears at the bottom of an e-mail message and may include the sender's contact information.

Some of the most common questions we hear are with regard to the signature file. Often-heard questions include: "How can I make my own?" "Why does yours have color and mine does not?" and "How did you do that?" The signature in question here usually appears at the bottom of an e-mail message and may include the sender's contact information.

From the **Tools** menu, select **Options...**. Click on the **Mail Format** tab. The bottom area contains the Signature information. You can create different signatures for new messages and also for replies or forwarded messages. Click the **Signatures...** button.

Create a signature by clicking **New**.... Enter a name to identify the particular signature. Then, in most cases, you will select the **Start with a blank signature** from the options, then click the **Next >** button.

From the Edit Signature dialogue, you can create the appropriate signature. Here you control the **Font** and colors along with **Paragraph** alignment. At this point, type in the text you would like to appear in the signature. You can also create more elaborate signatures by clicking the **Advanced Edit...** button to launch Microsoft FrontPage. Using FrontPage you can add graphics and more. Signatures are basically nothing more than miniature Web page files.

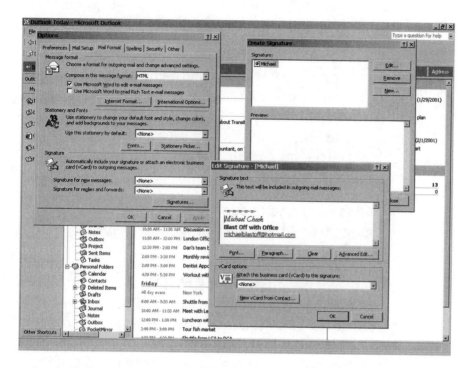

Figure 5.7

A basic signature need not be elaborate. Generally, enter a line or string of characters to break the signature from the body of the e-mail message. Your name and e-mail address is all you need after that (*Figure 5.7*). If you are creating a signature for a business account, you might want to add your title, the name and address of the business, and perhaps your office phone number.

A vCard can be added to your message as well. This attachment is a standard, virtual "business card," with all of the basic information. In order to create a vCard for yourself, you must have an entry with your pertinent information in the Contacts folder area of Outlook.

A vCard can be added to your message.

141

Attachments

The second most common type of question we hear deals with attachments. Attaching items to an e-mail message can be very easy. Look back at *Figure 5.2* and the e-mail message from Alan Greenspan, the first message listed at the top of the window. See the little paper clip next to it? That means Mr. Greenspan attached a file to the message.

If the message is opened, the attachment appears next to the field **Attachments:**.

Viruses, worms, and Trojan horses are types of malicious software that arrive attached to e-mail messages.

Before we go any further, we'd like to issue a warning about attachments. Be very careful! Viruses, worms, and Trojan horses are types of malicious software that arrive attached to e-mail messages. You might recall the *LoveBug* or *LoveLetter* from May 2000 or maybe the *Melissa* virus. All of these bugs arrived in e-mail messages. Don't think just because the e-mail arrives from someone you know that it's safe. Virus code often contains routines that automatically send e-mail containing copies of the virus to any or all addresses listed in Outlook. So, if a friend's computer gets infected, chances are you'll get an e-mail message from that friend with the virus attached. That's why we recommend that everyone install anti-virus software of some sort and keep it updated.

Now, since an e-mail message from Alan Greenspan was expected and anti-virus software is installed and updated, we can do three things:

1. Right-click on the attachment and select **Save As...** from the drop-down menu. Pay attention to where you save it, maybe save it to some place obvious like the Desktop.

2. Drag and drop the file somewhere. Select the attachment icon, press the left mouse button and hold it down, then drag the attachment to a location of your choosing.

3. Double-click and open the file. This option saves it to a temporary directory used by Internet applications. If you want to keep the file, you'll need to save it to a specific location on your hard drive. Please note that the temporary directory gets emptied, so your file will be erased if you don't store it someplace else.

Attaching a file is just about as easy. It can be done in three methods also.

1. When you create a new message, the paperclip icon appears near the top. Just click on it and find the file that you want to attach.

2. From the **Insert** menu item, select **File...**, and then find the file on your hard drive.

3. Drag and drop the file from its location into the new message window.

Within an organization, a form can be used to request time off or as a basic expense report.

Keeping you in forms

Within an organization, and using the server application Exchange, Outlook adds a few more features. One of these is forms. Some organizations use forms extensively. Within an organization, a form can be used to request time off or as a basic expense report.

Access a form from Outlook's **File** menu. Select **New** ▸ and then **Choose Form...**. The dialogue box pops up with the default entries. They'll all look familiar, like Messages, Contacts, Tasks, and Appointments. Click the drop-down field along the top that says **Standard**

The address book within Outlook can handle much more than addresses.

Forms Library. Your Personal Forms Library along with any corporate installations should be found there.

Designing a new form can be complex. The Microsoft Office Designer application is used to help create it. To explore creating a form, from the **Tools** menu, select **Forms ▼** and then **Design a Form…**.

Another handy feature with Outlook is called **Voting Buttons**. Like forms, this only works within an organization where everyone uses Outlook with Exchange Server. Say, for example, you're ordering lunch for your department. Everyone has the choice of chicken, beef, or vegetarian. You just need a count. Create a new message. In the message's toolbar, click the **Options…** button. The second area in the dialogue is dedicated to Voting and Tracking options. Click the box next to **Use voting buttons:**. The field next to it contains options like "Approve; Reject" or "Yes; No; Maybe." You can make your own, in this example, "Chicken; Beef; Vegetarian." Tell folks to click one of the choices, and it generates an automatic response.

You might be flooded with responses, but click only one. In the information area at the top, it says, "Click here to view the summary of responses." Do it to get a comprehensive list of what everyone wants for lunch.

Contacts

The address book within Outlook can handle much more than addresses. The basic interface looks something like *Figure 5.8*, with lots of details listed. Open or

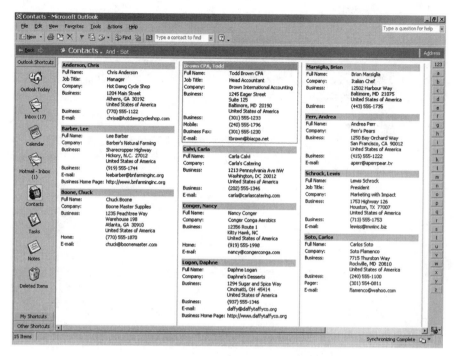

Figure 5.8

create a new record. Click on the **Details** tab, which offers a place for supplemental information like birthdays, spouse names, and more.

The **Activity** tab shows all the associations for that individual. You can create this linking automatically or manually.

Let's say you want to schedule a meeting. Just grab the contact entry and drag it to the **Calendar** icon in the **Outlook Shortcuts Bar**. Outlook creates the linking of that activity automatically. It's the same for e-mails and tasks.

To perform the linking manually, choose an appointment or a task and open it. A **Contacts...** button at the bottom will allow you to choose. Of course, if you invite

*The **Activity** tab shows all the associations for an individual.*

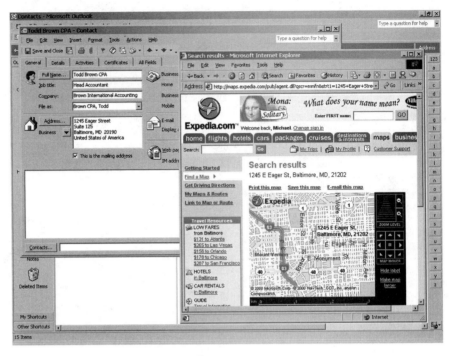

Figure 5.9

Outlook goes out and queries Expedia.com for a map of the location

the person to your meeting or assigned him or her a task, that person is included automatically. For e-mail, right-click any e-mail message you've received and select **Options...**. That includes another **Contacts...** button.

Back to the main Contact dialogue, notice the traffic sign icon in the toolbar named Display Map of Address. If you're connected to the Internet, click on it. Outlook goes out and queries Expedia.com for a map of the location, as seen in *Figure 5.9*. This can be a rather handy feature. Expedia.com can provide driving directions, too.

Microsoft has added instant messaging or IM addresses into the contacts fields. If you've got IM software installed, you can instantly chat with friends or coworkers. When you type something, it shows up at your contact's PC just moments after you hit **ENTER**.

New in Office XP

Instant messaging address field

Among the IM software, Microsoft has its own version. The Microsoft Network or MSN instant messaging software links directly into the contact. Therefore, when a contact is online with the MSN Messenger, a message appears in the window that tells you. This requires that you have MSN Messenger installed, too.

New in Office XP

Integrated support for MSN's instant messaging

The Calendar in Outlook works great in an office environment.

Scheduling

The Calendar in Outlook works great in an office environment. It might not be as handy in stand-alone mode, although certainly it can serve as a good reminder. Microsoft has been promoting an Internet standard for scheduling. But chances are, unless the recipient is using Outlook, too, scheduling remains something that works best within an organization with Exchange running at the server.

The Calendar icon in the Outlook Shortcuts Bar switches to the view of the current schedule, like in *Figure 5.10*. For users of previous versions, the calendar with Office XP looks a bit more colorful. Outlook now allows colors to be associated with different appointment types.

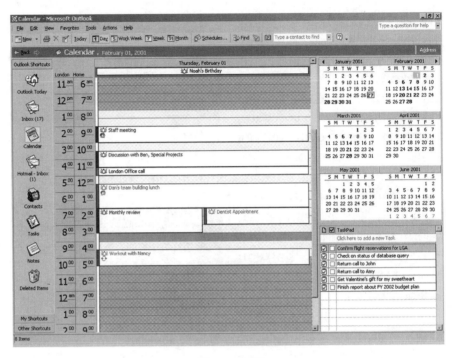

Figure 5.10

Outlook offers a new rule-based formatting that applies colors automatically.

New in Office XP

Colored appointments in the calendar

In previous versions, colors were used to denote free time, out of office, tentative, to busy. Now Outlook uses patterns to reflect that and has integrated more colors. By default, Outlook uses purple to denote vacation, green for personal appointments, red as important, blue for business, and several others. Some of the colors can be a little hard to read. Outlook calls these designators "labels."

Outlook allows the designators to change if you're just simply opposed to purple as vacation. The Calendar toolbar contains the **Calendar Coloring** button. Click on it and from the list at the bottom, select **Edit Labels...** to change what each color means.

Outlook also offers a new rule-based formatting that applies colors automatically. For example, dark yellow is associated with travel required. So if you're often taking a shuttle flight to New York, Boston, or

New in Office XP

Calendar rules to format colors with appointments

some other place, you might want to set up a rule that "shuttle" in the text always means a travel-required label (and therefore dark yellow color). You can also add a rule that when your boss sends a meeting, it's automatically colored red for important.

Using the same **Calendar Coloring** button, select **Automatic Formatting...** from the drop-down menu. Give the rule a **Name:** and select the **Label:** color. Then click the **Condition...** button to select the rules. For example, have it **Search for the word(s)** "shuttle" or apply colors based on who the meeting was **Organized By...**.

While Outlook can be very handy, let's be honest — sometimes Outlook can be a pest. Microsoft changed one of the attributes that annoyed us about previous versions of Outlook. If a few days passed while you were out of town on business and you returned,

New in Office XP

Unified reminder window to dismiss or "snooze" several notices at once

opening Outlook would result with several little windows popping up for every missed appointment.

You can add a rule that when your boss sends a meeting, it's automatically colored red for important.

Figure 5.11

When scheduling phone calls or coordinating other events, it's handy to know what time it is elsewhere.

No longer. Outlook now pops up a single window, like the 26 notices in *Figure 5.11*. Dismiss or snooze multiple notices at once.

When you received an appointment, Outlook offered three options — **A̲ccept**, **Te̲ntative** (for tentatively accepting), or **Decline**. Microsoft now offers a fourth option: **Propose̲ New Time**. This can be handy, especially when coordinating a large group.

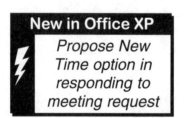

New in Office XP

Propose New Time option in responding to meeting request

Of course, Outlook still allows you to see schedules of people within a network. Microsoft also continues to support an Internet standard for scheduling for busy and free times.

In today's increasingly global economy, we've found that we need to schedule and relate with business and even personal contacts in other time zones. One of us works often with colleagues in London. When scheduling phone calls or coordinating other events, it's handy to know what time it is elsewhere.

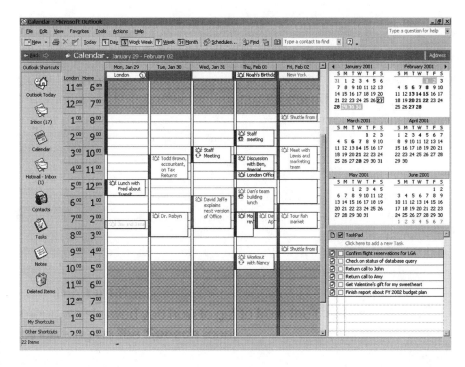

Figure 5.12

Outlook allows one other time zone to be represented in the Calendar, as seen in *Figure 5.12*. Go to the **Tools** menu, select **Options...** and then click the **Calendar Options...** button on the **Preferences** tab. The **Time Zone...** button in the bottom-right corner will allow you to show a second time zone and select it. Moreover, if you often travel to that time zone, click the **Swap Time Zones** button here to switch.

You'll also find a lot of handy features at **Calendar Options...**. If your workday begins at 10 a.m. and ends at 7 p.m., change that here. If you work Sunday through Thursday, you can set your workweek here, too. Outlook allows maximum flexibility in today's work world.

Outlook allows maximum flexibility in today's work world.

Tasks

Sometimes appointments are not appropriate for every item. For example, you've got to get that report done by Friday. Enter it as a Task rather than an appointment.

Tasks can have dates and times associated with them.

Tasks act like a To-Do list with some scheduling components. For example, let's say you've got that report due on Friday, but your colleague Juliana owes you some information. You can set a task for Juliana to provide you with her required information by Wednesday.

Tasks can have dates and times associated with them. A product like Microsoft Project offers more granular control over tasks and group projects. But Outlook can keep some basic information on track.

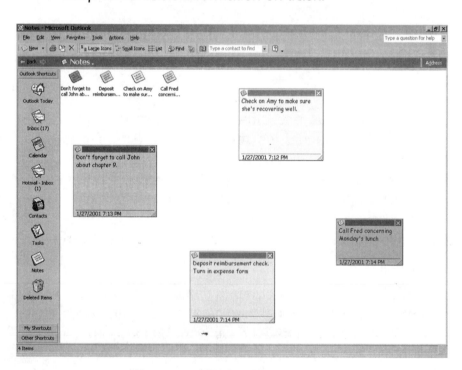

Figure 5.13

Notes

You know those little yellow sticky notes. Office has them too, as you can see in *Figure 5.13*. From the **Outlook Shortcuts Bar**, select the Notes icon. Click the **New** button in the toolbar and start typing.

To change the color of the note, just click the icon in the top left corner and select **Color ▼** and then choose one.

As long as Outlook is running, the notes will remain up on the screen. If Outlook closes, the notes disappear as well.

As long as Outlook is running, the notes will remain up on the screen.

6

PowerPoint

Putting Power into Presentations

PowerPoint can be the booster rocket for any presentation. We've observed how managers will flock to the person in the office who is good at using PowerPoint. It happens often in our office. More than one person has stopped by with disk in hand, containing a presentation using a plain black Times New Roman or Arial font on a white background. How boring!

PowerPoint can take the boring and blast it into orbit, especially with the new version included in Office XP. Office 2000 reorganized the presentation creation tool significantly, but Office XP has gone even further. Power users or those familiar with previous versions of PowerPoint will find this version radically different in both interface and in the features now available. In this version, Microsoft puts the power back in PowerPoint, and we are glad to see it.

PowerPoint can take the boring and blast it into orbit, especially with the new version included in Office XP.

Basic Concepts for Building a Presentation

The slides should show just the highlights found in the presentation.

Before we get too far into showing off PowerPoint, we thought it best to offer some basic advice on building a presentation. One important step should be accomplished before launching PowerPoint and that is the basic organization of the presentation. For example, think about how long your presentation needs to be, what is being discussed, and whether the presentation should be active or static (running from a computer, on paper, or overheads). These are all important considerations.

There are a couple more things to keep in mind when planning a presentation. If you are planning to print the presentation on an ordinary monochrome laser printer, it wouldn't make sense to use a lot of colors or dark backgrounds when creating it. If you are going to talk for 20 minutes, don't do 40 slides — that averages out to 30 seconds a slide, which is much too fast. We have found one slide for every two minutes is optimal, although up to 12 slides would be fine for that 20-minute speech.

A few more rules just to help you out:

♦ **Never read directly from a presentation**! The verbal portion of the presentation should express what the slides say in great detail. The slides should show just the highlights found in the presentation.

♦ Use images, clip art, etc., to express ideas. The old saying *a picture is worth a thousand words* holds true. Generally speaking, people tend to be more visually oriented rather than text-oriented. Use the presentation to illustrate your points.

♦ Organize your speech first, and then create the presentation around it. This process assures a more focused speech and slide show.

Now let's get on to the fun part!

The Interface

Launch PowerPoint and, from the start, it looks much different, especially for veteran users of PowerPoint 97. Just take a look at *Figure 6.1*.

New in Office XP

New, better-organized Interface

You can check out any of PowerPoint's 13 different toolbars by right-clicking anywhere in the toolbar area and selecting one.

Like all of the Office applications, the menus appear along the top. PowerPoint contains all of the standard entries with one addition: **Slide Show**. It also contains the same basic toolbars, including **Standard** and **Formatting**, which are both active in *Figure 6.1*. You can check out any of PowerPoint's 13 different toolbars by right-clicking anywhere in the toolbar area and selecting one. Or select the **View** menu and then the **Toolbars** command to choose one there.

Beneath the menu and toolbars, the left area of the screen is dedicated to two different ways of looking at the whole presentation: **Outline**

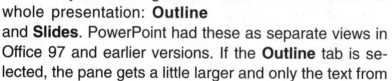

New in Office XP

Task Panes

and **Slides**. PowerPoint had these as separate views in Office 97 and earlier versions. If the **Outline** tab is selected, the pane gets a little larger and only the text from

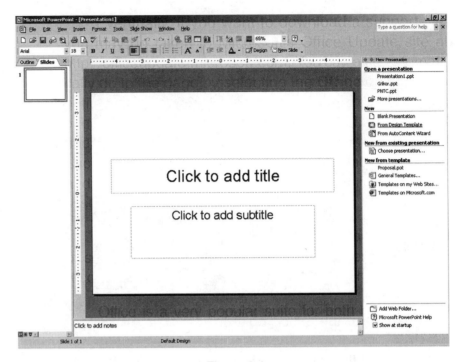

Figure 6.1

While all of Office now embraces Task Panes, PowerPoint is one application that can really take advantage of this new feature.

the presentation appears. If the **Slides** tab is selected, the pane is smaller and thumbnails of all slides appear. Slides can be moved around here, like in the **Slide Sorter** view, which we'll cover a bit later. The center of the screen is dedicated to the slide itself, of course. Just beneath the main slide is a small area for notes.

Perhaps the biggest change is the Task Pane area to the right. In this case, the Task Pane contains the basic information on opening an existing presentation file or creating a whole new one. While all of Office now embraces Task Panes, PowerPoint is one application that can really take advantage of this new feature.

Formatting a Presentation

When we launched PowerPoint, it politely created a blank presentation for us. Let's keep it for now and get some practice in by formatting it.

STEP 1: Select the **Fo̲rmat** menu and choose **Slide De̲sign...**. Notice next to the **Slide D̲esign...** item there's an icon that looks like a blue paintbrush over a plain white slide. You'll find that same icon next to the word **Des̲ign** in the **Formatting** toolbar.

STEP 2: The Task Pane changes. It now shows your current design, the designs you've used recently, and designs available. All appear in thumbnail images. If you place the cursor over one, you'll see its name. Find the one named "Proposal.dot" and right-click on it.

STEP 3: A menu pops up that gives you three choices: **Apply to A̲ll Slides**, **Apply to S̲elected Slide**, and **Show L̲arge Previews**. Since it is kind of difficult to see what the slide will actually look like, you can choose **Show L̲arge Previews** to get a better look.

STEP 4: Proposal.dot is the one we want. Left-click on it to select it. By default, that's like choosing **Apply to A̲ll Slides**, which we could have chosen in the last step where we decided to look at a larger preview first.

So now we have our first slide. Users of PowerPoint know this basic design will be applied to every slide we

When we launched PowerPoint, it politely created a blank presentation for us.

Master slides contain the design of the overall presentation.

create in this presentation. But this new PowerPoint has a trick. Notice in our menus, we had the choice of applying a design to selected slides. This is because PowerPoint 2002 can support multiple Masters.

New in Office XP

Support for mulitple master slide designs in a single presentation

Master slides contain the design of the overall presentation. In previous versions of PowerPoint, two slide masters were available: **Slide Master** and **Title Master**. You chose these for designing through the **View** menu, and it is still done that way in this version of PowerPoint, but with a slight change.

Select the **View** menu and choose **Master**. Master brings up three choices now, only one of them the **Slide Master**. Title Master is no longer here. Click on **Slide Master**.

The left pane changes and there is a new toolbar on the screen. The left pane contains two slides connected together by a bar. The top slide represents the regular slides for the presentation. The one beneath it is the title slide for the presentation (*Figure 6.2*).

Two kinds of slides exist, represented by slides 1 and 2. Regular slides occupy a majority of a presentation. The title slide is exactly that — the first slide you see in a presentation.

The **Slide Master View** toolbar offers a few new choices. The first is to **Insert a New Slide Master**. When you click on it, a new, blank slide appears. You can create a design yourself. Let's try creating one.

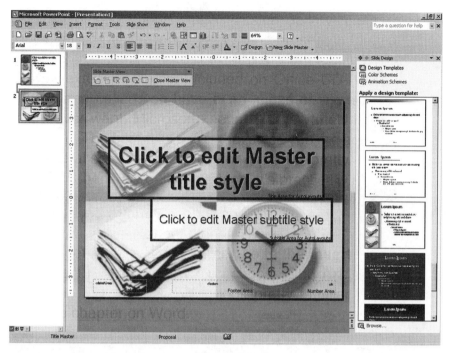

Figure 6.2

Notice the first slide with the column of images down the left side. Let's create a master with the column on the right.

STEP 1: In the left pane, click on the first slide. That takes you to that slide.

STEP 2: Click on the purple/yellow/green/blue image along the left side. Little dots and a box should appear around the image. Use the keyboard shortcut **CTRL+C** to copy it to the Windows Clipboard.

STEP 3: Now click on the third slide, the plain one created earlier.

STEP 4: Again using the keyboard shortcut, press **CTRL+V** to paste the image on that third slide.

Notice the first slide with the column of images down the left side.

STEP 5: Notice how the picture is over the text and that there is a shadow. It is really a box meant to fool us. Right-click on the picture and go to **Grouping** in the pop-up menu. Select **Ungroup**.

STEP 6: Click on the shadow area and hit the **Delete** key.

STEP 7: Now click on the purple/yellow/green/blue image and slide it all the way over to the right side.

STEP 8: Right-click on the image and select **Order** from the menu. Then choose **Send to Back** and the image goes beneath the text, as shown in *Figure 6.3*.

STEP 9: Resize the text boxes so they don't overlap onto the image and you are done.

Notice how the picture is over the text and that there is a shadow.

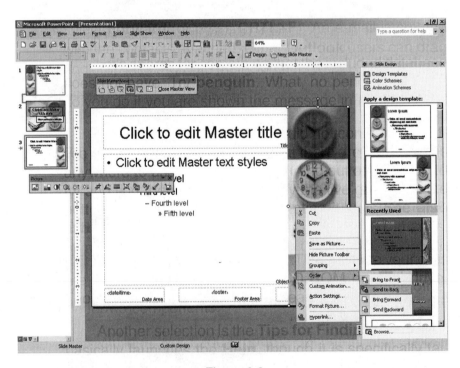

Figure 6.3

At this point, you should be beginning to understand a little about layers. As you become more familiar with PowerPoint, you can tap into the true power of this application by using layers.

Each element you create in a presentation occupies a vertical space. Of course, you can't see in three dimensions on a computer monitor. Think of it like the pages in this book. When the book is closed, page one is on top of page three, which is on top of page five, which is on top of page nine, etc.; these are layers. In PowerPoint, you can shift the layers. That is what we did in Step 8 — we moved the layer on top to the bottom.

Each element you create in a presentation occupies a vertical space.

Returning to the master slides, the second icon in the **Slide Master View** is **Insert New Title Master**. Title Masters are tied directly to the Slide Master, like the one we created earlier. This is why slides one and two appear connected. We don't need to create a new title master at this time.

The third button with the red "**X**" will delete a master. The fourth button with the yellow thumbtack is called **Preserve Master**, which allows PowerPoint to keep the Master slides you've created. You should notice there is a little thumbtack next to the slide three that we created.

The next button with the blue letter "**A**" allows you to name your master. The sixth button, called **Master Layout**, creates default text blocks. You already have the **Title**, **Text**, **Date**, **Slide number**, and **Footer**, so there is nothing to add. But, if you had deleted these placeholders, this button would restore them. Finally, there is the **Close Master View** button, which takes you back to slide creation.

In the master views, any changes to the basic format of the presentation can be created. For example, if you'd prefer a different font, change it here and it is reflected throughout the PowerPoint slides. You can create a template for all corporate presentations; make fonts larger or smaller; or add graphics or other content. Whatever is put on the masters appears on the other slides as each is created.

Alter the masters for practice before closing the example or perhaps change the fonts or font size. Or, you might want to change the bullets that PowerPoint chose. Right-click on a line on the plain slide master — the one that includes bullets. Choose any line. When the menu drops down, select **Bullets and Numbers...**. A window like the one in *Figure 6.4* will appear. You can choose any of the current bullets on that screen or select any character from any font. If you have a special bullet in a font like *Wingdings*, click on the **Customize...** button, select that font, and then choose the character. You are allowed to make it a percentage of the text size and can even choose different colors from the text.

PowerPoint 2002 also offers a new bullet offering. PowerPoint comes with several graphics that can be used as bullets or you can import your own. Of course, the image should be small. Click on the **Picture...** button. After a moment, a screen pops up with the selections available from Office. Notice the ones with a

> *In the master views, any changes to the basic format of the presentation can be created.*

New in Office XP

"Pictures" as bullets

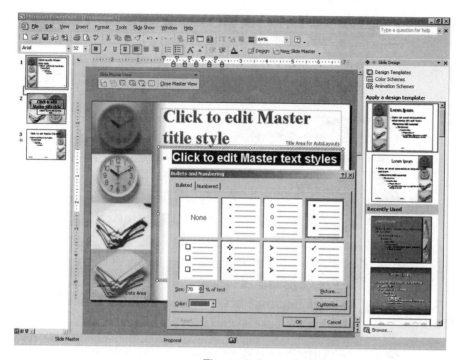

Figure 6.4

yellow star ☆ in the corner and some lines off to the right. This indicates the bullet appears to move or spin, or in some way is active. Choose any one of these or get your own by clicking the **Import...** button and finding an image on your hard drive, network, or some other storage device like a floppy, CD-ROM, or Zip disk.

From the **Bullets and Numbering** dialogue you can also choose the **Numbered** tab in order to create a numbered list. In addition to plain numbers (1, 2, 3...), it includes Roman numeral lists (i, ii, iii...) and letters (A, B, C...). You can customize your numbered list to fit the style of your presentation.

Now you can click the **Close Master View** button.

You can customize your numbered list to fit the style of your presentation.

Creating a Presentation

Now you should be back to the beginning of the presentation with the title screen. The two boxes on the slide read "**Click to add title**" and "**Click to add subtitle**." Let's add one now. Click on the title box and type something like, "**My Very First Presentation Using the New PowerPoint**."

The purpose of AutoFit is to reduce the font size so that all the text can fit into the box.

As you near the end of the title, you'll notice an icon off to the left probably appeared. That is PowerPoint's **AutoFit Smart Tag** (*Figure 6.5*).

We have a few choices. First, we can allow the font size to be reduced so that all the text will fit neatly into the box. The purpose of AutoFit is to reduce the font size so that all the text can fit into the box; this is the default setting. Second, while it's not included in the Smart Tag menu selections, we could allow PowerPoint to increase the size of the box to allow all the text to appear at full size. To do this, right-click the box and select **Format Placeholder...** from the menu list. You will also find **Placeholder...** in the **Format** menu. Click on the **Text Box** tab and then click the check box next to **Resize AutoShape to fit text**. Finally, we can simply rewrite the title so it fits a little better. In this example, let's shorten it up by changing the title to "**My First Presentation**."

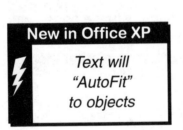

In the subtitle area, type your name.

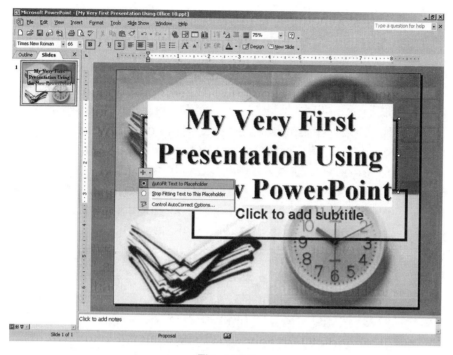

Figure 6.5

Now let's add another slide. The Formatting toolbar includes the button **New Slide**. If you don't see it among the choices, the **Insert** menu's first entry is **New Slide**, or you can press **CTRL+M**.

If the Task Pane disappeared earlier, it is back now with a selection of slide layouts. The top section, listed under Text Layouts, is just that — plain layouts with only text el-

ements. The next section of Content Layouts may have a title line, but includes areas for types of contents. Contents can include art, charts, movies, or tables (these choices are more thoroughly discussed in the next two sections of this chapter). The Text and Content section

*The Formatting toolbar includes the button **New Slide**.*

PowerPoint offers an impressive assortment of diagrams and organizational charts.

shows combinations of text areas with content areas. Finally, the Other Layouts area shows the "old" layouts that were included in previous versions of PowerPoint.

PowerPoint now handles the contents somewhat differently than in the past. Let's click one of the slides in the **Content Layouts** area. Let the cursor hover over the thumbnails for a moment until the name of the slide layout appears. Choose the one with the name "Content" — it is the one on the left in the second row. Just above the words "Click icon to add content" are six icons. In order, the icons each insert a different element:

◆ **Insert Table**: The table is like the grids found in Excel. Tables are excellent for explaining budget items or for otherwise organizing data. Of course, the size of the table and the amount of data can be limiting.

◆ **Insert Chart**: Charts include bar graphs, pie charts, etc. Use a chart to visually explain something mathematical, like how sales have gone up.

◆ **Insert Clip Art**: Office includes thousands of images of all sorts, not to mention thousands of more available from Microsoft's Design Gallery Live.

◆ **Insert Picture**: You can insert your own images in .gif, .jpg, .tif, or .bmp format.

◆ **Insert Diagram or Organizational Chart**: PowerPoint offers an impressive assortment of diagrams and organizational charts. There is more on this new feature a bit later.

◆ **Insert Media Clip**: Media clips include sound files (.wav, .midi, or .mp3), movie files (.mov, .avi, or .mpg), or other "active" elements such as an animated .gif.

Let's explore how to use all of these elements in a presentation.

Static Elements

Static elements lend themselves to any kind of presentation, from printed presentations or overheads to those straight from a PC to a projector. Static elements — or as Microsoft calls them, *content* — provide visual enhancements to an otherwise mundane presentation.

Tables

The grid pattern of a table helps organize information in a structured way. Users of PowerPoint 2000 will find the table process of PowerPoint 2002 very similar. But if you are jumping from Office 95 or 97 to XP, you will find the new PowerPoint to be much more robust.

Let's create a table element. In your presentation and just-created slide, click on the **Insert Table** icon. A box pops up to ask how many columns and how many rows. Let's make the table three columns and six rows. Then click **OK**.

The table fills up the whole rectangular area. Click inside any cell and the **Tables and Borders** toolbar appears (*Figure 6.6*). This toolbar looks and acts much like the **Table** menu in Word; the commands are very similar.

The picture icons in the toolbar provide quick access to common actions. You can add background colors to cells, draw borders between cells, merge multiple cells into one, split one cell into many, adjust vertical alignment, or evenly distribute rows or columns.

Static elements lend themselves to any kind of presentation.

Figure 6.6

The charting engine in PowerPoint taps into the exact same software code as found in Excel.

Clicking on the text **Ta̲ble** reveals all the other commands.

Charts

The charting engine in PowerPoint taps into the exact same software code as found in Excel. In fact, a miniature version of Excel pops up to help you enter data for the chart.

Let's create a new slide. Click on the **N̲ew Slide** button or press **CTRL+M**. Under **Content Layouts** choose the "Large Content" — the second slide on the first row. Click the second icon called **Insert Chart**.

PowerPoint pops up an already created bar chart with some sample data entered into the miniature Excel datasheet. You probably don't want the data already entered or you might want to create a different kind of chart. Let's make our own.

STEP 1: Highlight all of the data in the Excel cells.

STEP 2: Right-click and select **Delete...** from the pop-up menu.

STEP 3: In the box that pops up, select **Entire row...** then click **OK**. This clears out any formatting. The chart will disappear, but don't worry. We will get it back.

STEP 4: Right-click anywhere in the white area where the chart would be. A menu should pop up that includes the choice **Chart Type...**.

STEP 5: The familiar **Chart Type** dialogue box will appear, and, just as in Excel, you can choose the type of chart. How about a pie chart? In the left **Chart type:** box, click on **Pie**. In the right **Chart sub-type:** box, select the second one on the first row called **Pie with a 3-D visual effect**. Then click **OK**.

STEP 6: Now let's input some data. Notice how the miniature Excel datasheet has changed with the gray text like **Slice 1**, **Slice 2**, etc. Enter the names of the slices in those cells. Notice the row next to those cells does not have a number. Row 1 is actually the second row, meant to hold the actual data. Enter the following data about the population on a space outpost:

PowerPoint pops up an already created bar chart with some sample data entered into the miniature Excel chart.

Humans	Martians	Alpha Centurions	Venusians	Republicans
421	237	147	72	10

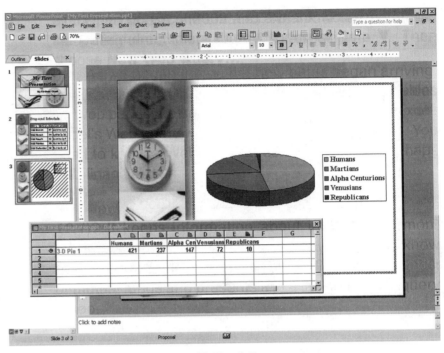

Figure 6.7

In this example, the species name belongs in the first row and the numbers belong in the second row, labeled row **1** (*Figure 6.7*). Enlarge the datasheet's window to see all of the entries.

STEP 7: Notice how PowerPoint generates a 3-D pie chart, as seen in *Figure 6.7*. It's a little thin, don't you think? We will take care of that. Close the miniature Excel datasheet by clicking on the **X** in the upper-right corner of the window.

STEP 8: Right-click anywhere in the white area around the chart and select **3-D Yiew...** from the menu. If this menu does not pop up, you may have right-clicked on the chart itself. Make sure to right click on a "blank" area.

If this menu does not pop up, you may have right-clicked on the chart itself.

STEP 9: The **3-D View** dialogue pops up. You can adjust **Elevation** angle so the pie appears a little fuller — in this case, perhaps 65 degrees. The pie can be turned by using the **Rotation** buttons. It can also be made to appear thicker by increasing the **Height** percentage of the base. When you're done adjusting it, click **OK**.

STEP 10: Change colors, fonts, and other options using the same method as in Excel. When you're done making changes, just click outside the chart area.

The chart is basically an OLE, or Object Linking and Embedding item. If you decide to edit it more later, simply double-click it.

Clip art

Microsoft includes thousands of clip art images on the CD-ROMs that come with Office XP. But Microsoft's larger gallery of images can be found on its Web site. That said, PowerPoint can insert clip art in two different ways.

Let's create a new slide. Click on the **New Slide** button or press **CTRL+M**. **Under Text and Content Layouts** choose the first one, "Text & Content." In the area to the right where "Click icon to add content" appears, click the third icon named **Insert Clip Art**.

The **Select Picture** dialogue appears with several pictures, as seen in *Figure 6.8*. You can type a word in the **Search text:** box and then click the **Search** button to find a particular type of image. For example, type in "computers" and then click **Search**. As you move the pointer

Microsoft includes thousands of clip art images on the CD-ROMs that come with Office XP.

Figure 6.8

over the images, a box pops up with the associated words describing each clip art image. The height and length size in pixels, actual file size, and the format of the clip art also appears. Most Microsoft clip art will be WMF format, which stands for Windows Meta File.

Find any image you like, click on it to select it, and then click **OK**. If you don't like it, just press the delete button and it will disappear. Then you can press the **Insert Clip Art** icon again and start over.

The **Picture** toolbar appears. The first icon, called **Insert Picture**, does just that. But we've already got a picture, so we can skip that.

The next group of icons can change the quality of the image. In order, they are **Color**, **More Contrast**,

Most Microsoft clip art will be WMF format, which stands for Windows Meta File.

Less Contrast, **More Brightness**, and **Less Bright-ness**. If you are printing out the presentation on a mono-chrome laser printer, it might be a good idea to press the **Color** icon to deselect it.

Four selections in a drop-down menu appear: **Auto-matic**, **Grayscale**, **Black and white**, and **Washout**. Selecting **Grayscale** turns a color image into an image of only black, white, and shades of gray. **Black and white** gets rid of the gray shades all together. **Washout** is good for using a faded image in the background; it makes the image appear very light.

Normally you cannot turn black-and-white or grayscale images to color because the color information is either not included with the picture or eliminated when it is changed to grayscale. However, selecting **Automatic** brings all the color back — this feature only works with Microsoft-provided clip art. If it is Microsoft clip art, the color will stay with the image even if you've selected **Black and white** or **Grayscale**.

The contrast and brightness buttons can increase or decrease these qualities. Sometimes you may want brighter, more vibrant colors, so increase the brightness and the contrast. Play around with this feature. Clicking on any of the "more" or "less" buttons will have an effect on the picture, though you may have to click multiple times for any significant change to occur.

Some of the more advanced tools, separated from the basic tools on the toolbar by a line, are **Crop**, **Ro-tate**, **Line Style**, **Compress Pictures**, and **Text Wrap**. Cropping lets you select a specific area within the photo to display, eliminating anything outside that selection.

Washout is good for using a faded image in the background; it makes the image appear very light.

Rotating lets you spin the photo — sideways, upside down, or anywhere in between. To rotate a photo, click **Rotate** and a series of dots will appear around the photo with one green dot. The green dot is the "hot dot." Grabbing it allows you to turn an image any increment of 360 degrees.

Line Style puts a border or frame around the image.

The **Compress Pictures** icon can reduce the "size" of an image. Every time you add a picture to a presentation, the file size gets larger. For example, let's say you add a high-resolution 8x10 picture to a document. You are going to use the photo at about 3x5 in size in the actual document. If you press this button, it reduces the file size so that it is 2MB instead of 10MB.

Also, some documents are not really designed to be printed out. For a photo to look good when you print the document out, the image needs to be at least 200 dots per inch (dpi), which is the default next to the **Print** selection in the **Compress Pictures** dialogue. Often, professionals in the field want at least 300 dpi. But, let's say you're going to make a Web page. When you click the **Compress Pictures** icon, you can select the **Web/Screen** radio button, which will reduce the dpi of the images in the document to 96 dpi.

The final important icon on the pane is **Text Wrapping**. This is helpful because the photo you imported does not have to stay in one place. You can grab it and drop it off anywhere in the document, even over text. The **Text**

Every time you add a picture to a presentation, the file size gets larger.

Wrapping icon will set up rules for how you want the text to behave and contour around the photograph, whether it should run straight though the image, form a box and remain outside of that image perimeter, or set as close to the photograph as possible.

The final button of the group, **Reset Picture**, returns the image to its original form. Let's say after all the adjustments, you want to start over. Click **Reset Picture** to do just that.

All of Office comes with clip art.

The other way you can insert clip art is the more traditional method found in other Office applications. The process is only slightly different and the Task Pane is a bit more powerful.

Go to the **Insert** menu and select the **Picture** command and then click on **Clip Art...** This opens the **Insert Clip Art** Task Pane along the right side of the screen.

All of Office comes with clip art. As in the other method, you can type a term in the Search text box and see if the Office collection of clip art has what you need. For example, when you type "computers" and push the **Search** button, you get the same examples as before (*Figure 6.9*).

New in Office XP

Automatic clip art search on the Design Gallery Live Web site at Microsoft.com

If you're connected to the Internet, Office even automatically goes and checks the Design Gallery Web site at Microsoft.com. Scroll down and you'll see a lot more image thumbnails

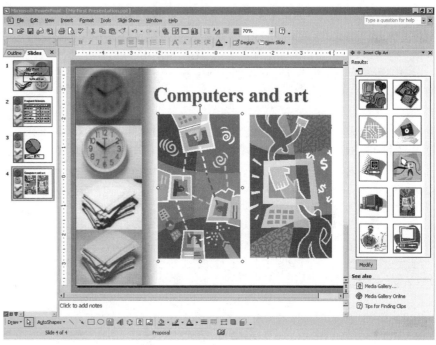

Figure 6.9

to choose from. Those from Microsoft's Web site have a small globe in the corner.

Choose one of the images. Click and drag it into the presentation or click on the image to bring up a menu on what to do with the image.

Enlarging clip art too much makes it appear blocky or pixeled.

Once you have highlighted the image, you can resize it by holding the cursor near the edge of the art until it turns into an arrow. Then holding down the left mouse button, move the mouse around to reduce or increase the image size. Be careful, though. Enlarging clip art too much makes it appear blocky or pixeled.

One particularly handy, but hidden, command can allow you to restyle clip art. If you double-click the clip art image, the Format Picture dialogue appears. Click on the

Picture tab. You should see a **Recolor...** button with most native, Microsoft clip art. Click that button.

A palate of colors appears. It shows all colors used in the art. You can selectively change some or all of the colors. If your primary company color is blue but a clip art image appears more green — the color of your company's primary competitor — change it here!

Pictures from files

Sometimes you want to put in an image of your own, whether it's a company logo or a photograph from your vacation. You can do it the same basic way as inserting clip art.

It's time to work on another new slide. Click on the **New Slide** button or press **CTRL+M**. Under **Content Layouts**, choose "Large Content." Where "Click icon to add content" appears, click the first icon in the second row — **Insert Picture**.

The window that appears now directs you to your hard drive. You now need to locate the file. PowerPoint can handle many different types of files:

♦ Windows bitmap and compressed bitmap files using the extensions .bmp, .dib, .rle, and .bmz.

♦ CorelDraw files using the .cdr extension.

♦ AutoCAD 2-D format using the .dfx extension.

♦ Micrografx Designer and Draw files using the .drw extension.

♦ Meta files including enhanced and Windows varieties using the following extensions: .emf, .wmf, .emz, .wmz, and .cgm.

You can selectively change some or all of the colors.

PowerPoint doesn't include a preview, so you need to know the filename when inserting your own image.

- Encapsulated PostScript files from art applications like Adobe Illustrator with the .eps extension.
- Graphics interchange format using the common .gif extension or the not-so-common .gfa.
- Joint Photographic Engineering Group standard images using the .jpg, .jpeg, .jpe, or .jfif extensions.
- Microsoft's own .mix files from PhotoDraw or Picture It! products.
- Kodak Photo CD images that use the .pcd extension.
- PC Paintbrush files using the .pcx extension.
- Macintosh image file format using the .pcz, .pict, or .pct extensions.
- Portable network graphics files with the .png extension.
- Targa files with the .tga extension.
- The often used Tag Image File Format for high-resolution images using either .tif or .tiff extensions.
- WordPerfect Graphics native clip art format using the .WPG extension.

That list may seem a bit overwhelming, but chances are you will use a .bmp, .gif, .jpg, or .tif file.

PowerPoint doesn't include a preview, so you need to know the filename when inserting your own image. Select the image file you want and click **Insert**. The picture is then placed into the slide. The image is automatically fitted to the space. You might notice a Smart Tag in one of the bottom corners. It comes up in case you don't want PowerPoint to automatically size the image to fit.

The **Picture** toolbar does not automatically appear like it does with clip art. You can get to it by right clicking in the toolbar area and selecting **Picture** from the drop-down menu.

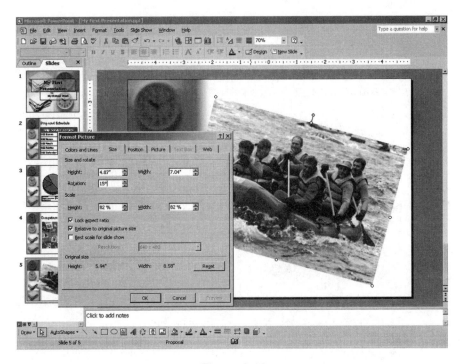

Figure 6.10

All of the commands remain the same. You can even choose grayscale and go back to the full color version if it's a Microsoft image. Office has become a lot more forgiving.

Double-clicking the picture brings up the **Format Picture**, just as in the clip art (although you cannot recolor your images). Format Picture controls offer a lot of options, even the ability to rotate an image (see *Figure 6.10*).

Everything is essentially the same as clip art. Go to the **Insert** menu and then select the **Picture** command. However, now click on **From File...**.

You can choose grayscale and go back to the full color version if it's a Microsoft image.

181

Diagrams and organizational charts

Microsoft added this powerful generation engine to all of the Office applications, but PowerPoint can really take advantage of diagrams and organizational charts.

Let's create another new slide. Click on the **New Slide** button or press **CTRL+M**. Under **Content Layouts** choose the "4 Pieces of Content" — the last selection.

After the organizational chart type is selected, PowerPoint generates a default chart with a manager and three subordinates.

New in Office XP

A whole new diagram and organizational chart tool

In any of the areas, you should click on the second icon in the second row to begin. The **Diagram Gallery** dialogue appears with six different types that can be created.

The first option, **Organization Chart**, was available in previous versions of Office. Microsoft made some changes here, though. All of the other types of charts — **Cycle Diagram**, **Radial Diagram**, **Pyramid Diagram**, **Venn Diagram**, and **Target Diagram** — are all new.

An **Organization Chart** — similar to a genealogy chart — sets up a hierarchical structure for businesses. Management is placed on top with subordinates and assistants falling beneath.

After the organizational chart type is selected, PowerPoint generates a default chart with a manager and three subordinates. The **Organization Chart** toolbar appears. This toolbar is different than the toolbar used with the rest of the diagrams. The first button — **Insert Shapes** — does just that, but only when a shape is selected. Just choose the manager for example, and you can add an **Assistant** or a **Subordinate**. Unfortunately, you can't have a two-manager shop, so a **Coworker** cannot be added here.

The **Layout** button helps create a style indicating how the subordinates *hang*. Small icons illustrate the style. You can play around with the different options.

The **Select** button can choose whole levels or groups in order to do some formatting, such as with colors.

Finally, the **AutoFormat** button that contains a cycle diagram and a lightning bolt provides some attractive formats. Click on it to see some options for design styles.

The **AutoFormat** button is available to all of the diagrams, too.

No matter the diagram chosen, the **Diagram** *toolbar stays basically the same.*

◆ A **Cycle Diagram** can be used to show a circular process, similar to the recycle symbol.

New in Office XP

Five new diagram types

◆ A **Radial Diagram** shows a relationship to a central item.

◆ A **Pyramid Diagram** illustrates the building blocks for some idea or process.

◆ The overlapping circles of a **Venn Diagram** show where processes meet.

◆ A **Target Diagram** helps focus the goals in on a bull's-eye.

No matter the diagram chosen, the **Diagram** toolbar stays basically the same. **Insert Shape** adds another item to the diagram — such as an overlapping circle for the Venn or an additional step for the cycle diagram (*Figure 6.11*).

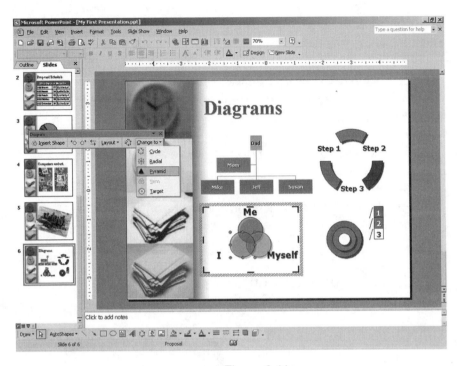

Figure 6.11

Any diagram can be changed to any other diagram type by clicking on **Change to** and selecting a different type.

Move Shape Backward or **Move Shape Forward** does just that. It moves that item around to different positions, like taking a bull's-eye and placing it the outer ring. If your diagram elements have different colors, it's easier to see this.

The **Layout** button just makes certain the diagram stays or extends beyond the confines of the block.

AutoFormat, as in the organizational chart, provides different designs.

Finally, any diagram can be changed to any other diagram type by clicking on **Change to** and selecting a different type.

Active Elements

Here is where PowerPoint's features start to really get fun. Active elements are the elements that *move* in some way. If you are creating a print-only presentation, this section doesn't really apply. You can't see a movie on a piece of paper!

However, if you're presenting using a computer, consider putting some action into your presentation.

The final element in the content selection is **Insert Media Clip**. A media clip can refer to a variety of items.

♦ Pictures and Photos: Like the **Insert Clip Art** and **Insert Picture**, the media clip supports all of those types of files. Please note among them is the Graphics Interchange Format or .gif. This kind of image has the ability to be animated. While doing a presentation, the animated .gif can appear to move, as it should. PowerPoint 97 and previous versions did not support the animation in .gif files.

♦ Sounds: Like images, PowerPoint now supports 12 different file extensions, including .aif, .au, .midi, .mp3 and .wav. Chances are good that any sound file will work.

♦ Motion: Think of these as movies. In addition to the previously mentioned .gif file, PowerPoint can import plenty of other file types, including .mpeg, .avi, Real Media's .ram or .rm, Apple QuickTime's .qt or .mov, and several others.

Let's experiment a bit. Click on the **New Slide** button or press **CTRL+M**. Under **Content Layouts** choose the "2 Small and 1 Large Piece of Content."

Active elements are the elements that move in some way.

Figure 6.12

*You can click on the **Insert Media Clip** icon to open the **Media Clip** box, similar to the **Clip Art** dialogue.*

You can click on the **Insert Media Clip** icon to open the **Media Clip** box, similar to the **Clip Art** dialogue. If you choose a sound and click **OK**, you'll be asked whether to play the sound automatically or only when the icon is clicked. You can decide when the sound should play a little later, but go ahead and click **Yes** for now.

An icon that looks like a little speaker appears. This is the embedded sound. However, you don't always want people to be able to see the sound, just hear it.

Here are a couple of tricks to hide the icon. One option is to drag the icon just off to the side, beyond the boundaries of the slide, as seen in *Figure 6.12*. The sound still plays on cue.

The other option is to cover it by putting an image or a block of the same color as the background over it. This goes back to the concept of layering we talked about earlier in this chapter.

Do you have a sound on the hard drive that might be of use? Import it into the **Media Clip** gallery. From the Media Clip dialogue, click the **Import...** button on the bottom left, find it on the hard drive, and click **Add**.

Microsoft offers several sounds and movies for use from its Web site.

Like clip art, Microsoft offers several sounds and movies for use from its Web site. Select one of the "Click icon to add content" areas, but do not choose an icon. From the **Insert** menu, choose **Movies and Sounds**. Both **Movie from Media Gallery...** and **Sound from Media Gallery...** look for content included with Office, either on the hard drive or Office CD-ROM, if it's in the CD-ROM drive. If you're connected to the Internet, it will also pull up sounds or movies from Microsoft.com's Media Gallery Online Web site.

Choosing either **Movie from File...** or **Sound from File...** looks only at the hard drive or the local area network.

New in Office XP

"Play" an individual slide to see animated content in action

Insert a few different clips into your presentation. If you insert a sound or a movie, check out the current slide you are working on in the left panel with the thumbnails of the presentation slides. You should see a tiny yellow star just to the left of your slide. Clicking on the star will play the individual slide.

In previous versions of PowerPoint, tiny thumbnails would show animations. Otherwise, the entire slide show would have to be launched. This new handy shortcut provides a quick view of what a slide will do. If the slide animation is in process and you want it stopped before it's done, just hit **ESC**.

A picture, chart, words, or anything else that you can click on, can appear, move, or disappear in a variety of ways.

Animating the Whole Show

Besides active elements such as movies and sounds, PowerPoint provides impressive animation effects for all the elements that don't appear so active.

Save your presentation and let's go back to the beginning, to the first slide.

Animation has always been a part of PowerPoint. With each generation, Microsoft has made the animation more sophisticated and smooth. If your PC offers a decent graphics accelerator, like an Accelerated Graphics Port (AGP) card with at least 16 megabytes of video memory, PowerPoint can provide a little extra oomph in those presentations.

Animation in PowerPoint basically takes two different forms: individual objects and slide transitions. A picture, chart, words, or anything else that you can click on, can appear, move, or disappear in a variety of ways. That is how the individual object animation works.

Slide transition is another kind of animation; it switches from one slide to the next — from slide one to slide two, two to three, three to four, and so on.

Let's begin with slide transitions. You should be back at slide one. From the **Sli̲de Show** menu, select **Slide**

<u>T</u>ransition…. The **Slide Transition** Task Pane opens.

Many choices appear — now 57, up from the 40 transitions available in Power-Point 2000. The area near the top of

New in Office XP

17 new slide transitions

the Task Pane is headed by **Apply to selected slides:** followed by selections like Box Out, Checkerboard Across, Cut Through Black, Newsflash, Shape Circle, Uncover Down, Wedge, and Random Transition.

Skip Random Transition for now and start clicking on some of the selections. As you click each one, some animation should appear. If not, at the bottom of the Task Pane make sure the box next to **AutoPreview** is checked.

Choose any one, perhaps the "Wheel Clockwise, 4 Spokes" like the one used in *Figure 6.13*. In the Task Pane beneath the different slide transition selections, there is an option called **Modify Transitions**. Change the **Speed:** between Fast, Medium, and Slow. **Sound:** can also be added to the transition. Only .wav files can be used here. We'll show you later how to add a soundtrack to your presentation.

Beneath that is the **Advance Slide** area. This determines whether you want to go to the next slide by the click of the mouse or after a delay of a specified amount of time. If you're going to be presenting in person, you might want to click to advance to the next slide. If the slide show is unmanned or timed, select **Automatically After** and specify a time, usually in *minutes:seconds* format. If you type in *5*, PowerPoint assumes you mean 5 seconds.

Skip Random Transition for now and start clicking on some of the selections.

Figure 6.13

The ***Play*** *button shows what the current animation looks like in the current slide.*

You can also select both click and automatic advance. If this option is used, PowerPoint moves to the next slide based on whichever comes first — the click or the elapsed time.

The buttons at the bottom, **Apply to Master** and **Apply to All Slides**, apply whatever slide transition you have selected to be used. **Apply to Master** assures all future slides have the same transition. **Apply to All Slides** puts the transition into play on all existing slides but not future ones.

The **Play** button shows what the current animation looks like in the current slide. **Slide Show** begins playing the slides from the current slide. For example, if you are working on Slide four, the **Slide Show** will begin at slide four.

If you thought transitions were a cool feature, just wait for individual-object animation!

In previous versions of PowerPoint, each item needed to be selected to apply animation. With Office XP, PowerPoint

offers **Animation Schemes**. These are kinds of preprogrammed animations for the common or default elements in a presentation.

Go to slide one. From the **Slide Show** menu, select **Animation Schemes…**. The different schemes appear in a Task Pane. These schemes range from **Subtle** effects, where the title and subtitle fade into place, to **Exciting** effects where text flies, flashes, or explodes onto the screen. Some **Moderate** effects balance out the two. Click on any one to see an example. If nothing happens when you click a scheme, make sure **AutoPreview** at the bottom of the Task Pane is checked.

Like transitions, you can choose to apply these schemes to all slides or even to the master slides. But let's not do that. Choose one scheme and move to slide two.

Right-click the title on this slide and select **Custom Animation…** from the drop-down menu. The **Custom Animation** Task Pane appears.

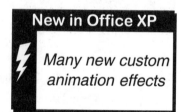

PowerPoint 2002 offers so many new animation effects; it takes a while to count them all. The previous version included 17 different effects with settings that

If you thought transitions were a cool feature, just wait for individual-object animation!

191

gave a grand total of 54. For example, the animation "Fly-In" offered settings for From Bottom, From Top, etc. — basically eight directions, so eight different effects.

With all of the different settings for each effect, there are now hundreds of combinations.

PowerPoint 2002 starts with those same 17 effects and adds 35 more. With all of the different settings for each effect, there are now hundreds of combinations. With so many animations available, it can be a little overwhelming.

New in Office XP

Several animations can be added to a single object

And speaking of overwhelming, Microsoft has added a lot more flexibility to Power-Point. In previous versions, each object could have only one animation effect. Now each object can have more than one effect applied to it.

And there is more!

New in Office XP

Objects can have multiple animations for entrance, emphasis, and exit

PowerPoint now offers additional animation types. Animations for the **Entrance** of the object or how it first gets onto the slide are included. That has been the standard for PowerPoint.

Now PowerPoint offers animations to highlight — known as **Emphasis** — while the object is on-screen. Animation for the **Exit** of the object, or when it leaves the slide, is also included. Those two new items add 76 more animation effects without counting the different settings. And we're not done yet!

PowerPoint now offers Motion Paths — paths that the object moves along. Power-Point offers 66 of them or you can even design a custom path.

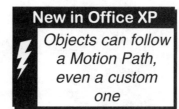

New in Office XP

Objects can follow a Motion Path, even a custom one

*You can put multiple **Entrance** or **Exit** animations on the same object.*

If you add all the basic animations, you have up to 194 different effects with which to start. After applying all the various settings, PowerPoint offers thousands of options. Fortunately, the interface is manageable.

So let's take a look at the **Custom Animation** Task Pane again. The top left button — **Add Effect ▸** — begins the process. When clicked, the drop-down menu offers **Entrance**, **Emphasis**, **Exit**, and **Motion Paths**.

As mentioned before, you can apply more than one animation to any object. Yes, in fact, you can put multiple **Entrance** or **Exit** animations on the same object. But for our exercise, we will have the title enter, provide a bit of an emphasis, and then exit simply and smoothly.

STEP 1: Click **Add Effect ▸** and choose **Entrance** from the drop-down menu. Another menu appears with several choices, but click on the bottom one, **More Effects....**

STEP 2: The **Add Entrance Effects** window appears. Microsoft divides animation effects into **Basic**, **Subtle**, **Moderate**, and **Exciting**. Click on a few of them. If the **Preview Effect** box at the bottom of the window is checked, you can see what the animation will look like (*Figure 6.14*). Now choose one. How about **Grow & Turn** under the Moderate group? Now click **OK**.

Figure 6.14

Microsoft offers five different settings — Very Slow, Slow, Medium, Fast, and Very Fast.

STEP 3: The heading on the Task Pane has changed to **Modify: Grow & Turn**. The **Start:** box has a little mouse icon (🖰) with the words "On Click." That means the animation will not begin until the mouse is clicked. Click the down arrow for the drop-down menu and select "With Previous" instead. That causes the animation to begin automatically. The **Property:** box is blanked out because **Grow & Turn** doesn't offer any additional settings. But you can change the pace at which the **Grow & Turn** occurs using the **Speed:** box. Here Microsoft offers five different settings — Very Slow, Slow, Medium, Fast, and Very Fast. Select one to see the change. Now we are basically done with the **Entrance**.

194

STEP 4: Click **Add Effect** ▸ and choose **Emphasis** from the drop-down menu. Again choose **More Effects...**.

STEP 5: The **Emphasis** effects appear in three areas — Basic, Subtle, and Exciting. In the Basic area choose **Spin** and click **OK**.

STEP 6: Now the Task Pane should reflect the new effect selection with **Modify: Spin** at the top. Again, we've got the **Start:** ⬚ "On Click" choice. Let's change that again. Now there's an additional choice — either **With Previous** or ⊕ **After Previous**. **With Previous** executes the animation in concurrence with the animated object above it. If you select **With Previous**, it wouldn't work since it's the same object. Choose ⊕ **After Previous** so the animation begins immediately following the other animation. The **Amount:** box can put the degree of the spin — how far and which direction. You can see the choices, but leave it at **Full Spin** and **Clockwise**. And as before, choose the **Speed:** at **Medium**.

New in Office XP

Multiple animations can occur at the same time

*The **Amount:** box can put the degree of the spin.*

STEP 7: Click **Add Effect** ▸ and choose **Exit** from the drop-down menu. Again choose **More Effects...**.

STEP 8: The **Exit** effects appear generally like the **Entrance** effects in four areas: Basic, Subtle, Medium, and Exciting. Under the Subtle area, choose **Faded Swivel** and click **OK**.

STEP 9: Like before, pick ⊕ **After Previous** and **Medium** speed. However, this time, we want a delay before the **Exit** animation begins. To the right side of the Exit animation entry in the Task Pane is a down arrow. Click it and from the drop-down menu, select **Timing…**.

STEP 10: In this dialogue, you can choose the amount of delay in seconds. Let's boost it to five seconds. Now click **OK**. Click the ▸ **Play** button in the Task Pane to see the entire sequence of animation together.

With PowerPoint offering so many options, the animation can really become complex.

With PowerPoint offering so many options, the animation can really become complex. Of course, all the objects in a presentation can be animated. It is important to remember that each animation taxes your processor a bit. Presentations that are graphically intense could run slowly on some laptop computers.

The Task Pane puts the items in order of the animation. The ⇧ **Re-Order** ⇩ with arrow buttons at the bottom of the Task Pane shift those objects into the desired order.

Go to slide 2 and select the pie chart we created. The **Custom Animation** Task Pane is still probably up on-screen. If not, right-click the pie chart and select **Custom Animation…** from the drop-down menu.

Choose an **Entrance** effect for the pie chart — maybe the **Fade** animation under the **Subtle** area. When ▸ **Play** is pressed in the Task Pane, you'll see the entire pie chart appear.

In the object's drop-down menu within the Task Pane, select **Effect Options…**. The **Fade** dialogue box appears. Click on the **Chart Animation** tab and change the **Group chart:** entry from **As one object** to **By category**. Click **OK**.

Now press ▶ **Play** in the Task Pane. Notice how the pie appears one slice at a time. PowerPoint allowed charts to be built in previous versions, but PowerPoint 2002 is very different in accessing that feature.

Notice the bar beneath the object's entry in the Task Pane. See the small double arrows in that bar? Click them and the five "categories" appear. Here you can change the animation for each individual slice if you prefer.

A similar breakdown is available for diagrams like the ones in slide 6. Animation can be complicated, but it is a feature that's fun. So, play around with it until you get the hang of it.

Animation can be complicated, but it is a feature that's fun.

Special Features

PowerPoint now makes presentation creation much easier since Microsoft provides almost completely packaged presentations with the basic content ready to go.

Close out of the current presentation we've been working on. From the **File** menu, select **New** or press **CTRL+N**. The **New Presentation** Task Pane appears. Choose **From AutoContent Wizard** under the **New** area.

New in Office XP

New AutoContent Wizard

The AutoContent Wizard launches. Click **Next >** to get started. Microsoft has organized 23 different presentations into five categories: **General**, **Corporate**, **Projects**, **Sales/Marketing**, and **Carnegie Coach**.

PowerPoint generates a seven-slide presentation with placeholders for all the content.

Choose one from the selection that looks interesting. How about **Communicating Bad News** from the **General** category? Now click **Next >**. The next question asks how the presentation will be used: **On-screen presentation**, **Web presentation**, **Black and white overheads**, **Color overheads**, or **35mm slides**. As mentioned before, you need to determine where a presentation will be used in order to make sure it's best suited for that output.

The overheads obviously print to a printer while the 35mm slide option requires sending the file to a special printer or a service. Choose **On-screen presentation**. Click **Next >**.

Choose options here. What is the title of the presentation? Make one up and add a Footer. Click **Next >** and you will see you could have just clicked **Finish**.

PowerPoint generates a seven-slide presentation with placeholders for all the content. Notice the placeholders offer advice like, "Key points to remember that will give audience confidence or improve morale."

Don't like the template design? Change it! From the **Format** menu, choose **Slide Design...**. You should recognize the **Slide Design** Task Pane. Choose another template. Add slides, clip art, sounds, and more. Some content may be animated, but you can spruce that up, too.

Set up a show

Different kinds of on-screen presentations need different kinds of setups. Do you want the presenta-

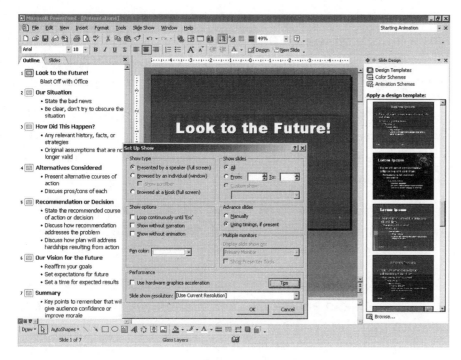

Figure 6.15

tion to run continuously? Are you presenting this with narration?

Under the **Sli̲de Show** menu, select **S̲et-Up Show…** to see some of the options available (*Figure 6.15*). For example, if the presentation is to show at a kiosk, perhaps at a trade or expo show, you won't be there to click to move through the slides. So choose **Browsed at a kiosk (full screen)**.

When setting up a show, you might want a custom version based on the appropriate client. Under the **Sli̲de Show** menu, pick **Custom Sho̲ws….** Here you can create different shows. With custom shows you can skip, repeat, or reorder slides.

When setting up a show, you might want a custom version based on the appropriate client.

To make a slide show more interactive, put in some **Action Buttons**. This command is also found under the **Slide Show** menu. PowerPoint provides some standard button designs like one for "home" to return to the first slide.

You don't have to put in an **Action Button** to create an interactive presentation. Any object can have an associated action. Choose an object and right-click. From the menu, select **Action Settings....**

Action Settings can be powerful. The **Mouse Click** tab should be selected. That means whenever the mouse clicks the selected object, the specified task will occur.

Hyperlink to: can maneuver around a slide show, launch into another PowerPoint presentation file, or find a particular slide within that presentation. **Hyperlink to:** can even connect to a particular Web site if the PC is connected to the Internet.

Other options include running a particular application or launching a macro. PowerPoint can also play a particular sound.

The **Mouse Over** tab offers the same basic options, but causes the action to occur from moving the cursor over the selected object.

Other Tips and Tricks

PowerPoint can seem a little daunting, but don't let it be. Play around with it; become familiar with the options it offers. Here are some more tips and tricks to help you along.

PowerPoint can seem a little daunting, but don't let it be.

Advancing (and backing up) through a slide show

You don't have to use a mouse to move from slide to slide. The keyboard is just as handy.

Press the **PAGE DOWN**, **SPACE BAR**, **ENTER**, or the **N** key (for "next") to advance during a slide show. To back up, press **PAGE UP**, **BACKSPACE**, or the **P** key (for "previous") to move backward.

Color schemes

When you select a template design for a presentation, PowerPoint uses a palate of complementary hues known as a **Color Scheme**. Each scheme is eight colors. By selecting a color not in the scheme (by using **More Colors...** from the palate choices), PowerPoint adds colors to the scheme. Sometimes though, the palate just isn't right or appears a little off. You can select a whole new Color Scheme for the presentation.

From the **Format** menu, select **Slide Design...**. When the **Slide Design** Task Pane appears, click on **Color Schemes** at the top. Several built-in schemes appear in the Task Pane. Select one or click **Edit Color Schemes...** at the bottom (*Figure 6.16*). Here you can choose the tint you feel is just right for the text, background, shadows, fills, and accents.

Soundtracks

Soundtracks to presentations provide a little pizzazz, especially for kiosk or unmanned presentations.

You can select a whole new Color Scheme for the presentation.

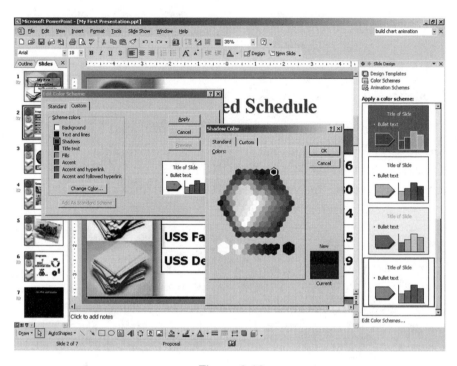

Figure 6.16

Avoid .wav files since these sound files are very large if they are longer in length.

STEP 1: From the **Insert** menu, select **Movies and Sounds** and then **Sound from File...**. Choose the sound file you want to use — most likely a .midi or .mp3. Avoid .wav files since these sound files are very large if they are longer in length. You want something that plays for a long time.

STEP 2: When it asks if the sound should play automatically, click **Yes**.

STEP 3: Now drag the little icon that appeared when you added the sound file off to the side so it won't appear in a slide.

STEP 4: Right-click the sound icon and go to **Edit Sound Object**. It is here where you decide whether or not the sound should be looped, which means it repeats itself until stopped. Click the box next to **Loop until stopped** if you want it to repeat.

STEP 5: Right-click and select **Custom Animation...** from the menu. The **Custom Animation** Task Pane appears. You will probably want the sound to be the first animated object. Click the **Re-Order** arrows at the bottom of the Task Pane to put the sound in the first position, known as position 0 (zero).

STEP 6: In the Task Pane, click the down arrow **6** to the right of the sound at position 0. Select **Effect Options...** from the drop-down menu. In the Stop playing area, click one of the appropriate radio buttons — probably the **After: *X* slides**, where *X* is a number. If the sound is supposed to play throughout the presentation, choose the number of slides in the presentation. Hitting **ESC** ends the presentation and stops the sound. You might also want to set the **Sound volume:** in this dialogue box. Click **OK** when you're done and you've got a soundtrack!

Once you have a complete presentation, printing it out on paper can sometimes be a task.

Printing the Presentation

Once you have a complete presentation, printing it out on paper can sometimes be a task. Microsoft has improved this feature over the last few versions. In the past, it could sometimes take hours to print out a full presentation. It can still take some time. Realize that PowerPoint incorporates some complicated graphics that can be difficult for any printer to handle.

Microsoft now includes **Print Preview** in PowerPoint. Microsoft has designed it with some intelligent features.

For example, if you are printing to a monochrome laser printer, PowerPoint knows to generate the preview in grayscale black and white.

Print Preview is found under the **File** menu. But it can also be accessed from the **Print...** dialogue. Press **CTRL+P** and the **Print...** dialogue box appears.

Sometimes the computer used in a presentation isn't your own.

Let's say you're going to give an on-screen presentation, but want to provide a copy of the slides to your audience. In the **Print...** dialogue, select **Handouts** under the **Print what:** area. The **Handouts** area is highlighted and you can select how many slides you want per page and other options. Then click the **Preview** button and see what the handouts will look like.

Pack and Go

Sometimes the computer used in a presentation isn't your own. You need to get a presentation from your PC to another PC and the Internet isn't an option. **Pack and Go...** feature provides the solution.

Under the **File** menu, select **Pack and Go...**. It launches the **Pack and Go Wizard**. Click **Next >** to get started. If you are in the presentation you want to pack, check **Active Presentation**. You can add other files by selecting **Other Presentation(s)** and picking them out. To choose multiple files, hold down the **CTRL** key while left-clicking on each. Click on the **Next >** button now.

The next step asks for where the file should be written. You can write to the A:\ floppy drive. If the presentation is larger than 1.4 megabytes, don't worry. **Pack and Go** will split it up over multiple floppy disks and reassemble it on the destination system. **Pack and Go** can also write to virtually anywhere, including Zip disks. Choose a destination here then click the **Next >** button.

Now you must decide whether to include linked files and to embed the fonts. You probably want to include the linked files like graphics, movies, and sounds, so check **Include linked files**. If the presentation uses unusual fonts that are not likely to be included on the destination system, check the **Embed TrueType Fonts**. Otherwise you can skip that. Click **Next >.**

Microsoft offers a small executable file for systems that don't have PowerPoint installed. You can choose to have the viewer included in the packed file. You can also download the viewer by clicking the **Download the Viewer** button. Click **Next >** and then **Finish** to save the compressed file out to disk.

That's it; you are done. It is time to go and present a powerful, interesting, and creative PowerPoint presentation!

Pack and Go will split it up over multiple floppy disks and reassemble it on the destination system.

7

Access

If you have ever seen those movies where the space hero lands on a strange, alien planet and then has to learn how to survive, you might understand a bit of what a first encounter with Access might be like.

Of all the components in Office XP, Access is probably the most powerful and the most complex. The spreadsheet component of Excel has been increased in functionality to a point where it can perform many of the same duties as Access. If you are thinking about very basic database-type applications, you might consider using Excel instead. Don't try to kill a mosquito with a cannon!

Access has a history of being a solid database. It has fought a lot of competition over the years with applications like FileMaker Pro, Lotus Approach, and Corel Paradox. Access has a lot of battle scars and patches to prove it. The version of Access that comes with Office XP is relatively easy to use, though at first it may seem like the application is in a foreign language. Many terms and

Access has a history of being a solid database.

functions need to be learned in order to tap into the features of Access.

Though we concentrate on the new and improved features in this guide, we will also try to explain some of the basic terms and functions for those who have never tackled a database before.

Access can take boxes that contain a lot of information and pull them together in a whole new way.

Basic Concepts in Databases

Think of a database as a giant box that can hold many items. You can put just about anything into this box and the box grows larger as needed to accommodate new items.

If you are familiar with Excel, it probably sounds a bit like its explanation — boxes that can contain information. However, Access is much more powerful. Access can take boxes that contain a lot of information and pull them together in a whole new way.

Access offers many special features that no simple box could ever hope to match. For one, you can automatically pull an object — or a group of objects with similar properties — to the top of the box. You can also assign special properties to each item, such as schedules for maintenance, and have those objects automatically pulled out as needed.

The only restriction on how large a database can be is how much space is available on a computer and, to an extent, how fast the system can crunch data requests.

As an example, I know of one extremely detailed database a friend created for a coal mining company. The database contained everything the company did, from wages to roofing bolts in the mine to paper clips used at the home office. Even though the database was running on a gigahertz computer specifically designed for databases, a simple query still takes upward of five minutes to complete. It might take hours on a slower system.

Again, make sure you need the power of Access before attempting to create a database. Access can handle small datasets just as easily as large ones, but why go to through the complexity when other applications, such as Excel, can handle small jobs just as easily?

If you are still reading this, then you either have a large amount of data you would like to track, or you have used Access in the past and plan on using it in the future.

Those of you with existing Access databases will be pleased to know that they are completely compatible with the version in Office XP.

Those of you with existing Access databases will be pleased to know that they are completely compatible with the version in Office XP. If you are using the database found in Office 2000, you won't even need to convert data to the new format. Office XP offers a way to edit and modify Office 2000 Access files without conversion. If you choose, you can convert the files anyway.

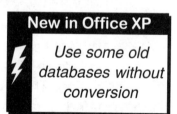

New in Office XP

Use some old databases without conversion

If you are using an Access database from Office 95 or 97, however, you will need to undergo a conversion process. If all goes smoothly, this process is more or less automatic. If something goes wrong, however, you will be given a detailed chart with information about each and every error that occurred. Simply follow each error and you should be able to manually troubleshoot problems on a case-by-case basis.

Mostly this means you need to reenter some data, though in detailed tests, 99.5 percent of all conversions went smoothly, even with very large databases.

The Database Wizard lists fields and tables that will be in the database.

Creating a database

If you are starting from scratch and have never used Access or any other database program before, a good idea is to see if any of the preexisting templates that come with Office will work for you.

STEP 1: Select **New Office Document** from the Windows Start menu and then click on the **Databases** tab. All database templates are stored here. Pick one like **Expenses** to load up a template.

STEP 2: You'll be asked to name the file where the database will be stored first. You can leave the default name, **expenses.mdb**.

STEP 3: Access then loads the template. The Database Wizard lists fields and tables that will be in the database. The information to the left lists the tables. Think of tables as a collection of related data. In the expenses database, four tables are available: **Information about employees**, **Expense report information**, **Expense details**, and **Expense categories**. Click on each to see the related fields for each table. The list on the right has the fields.

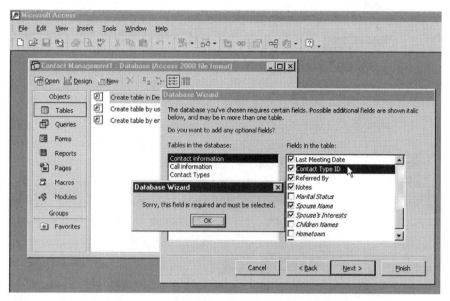

Figure 7.1

Some of these fields will be required, meaning that each must remain in the database. These fields appear in normal text and have boxes next to each that are already checked. You can't uncheck these; if you try, you will get a very apologetic reply from Access saying that doing so is beyond its control, as seen in *Figure 7.1*.

Some of the fields will be written in italics. These can be added to the database if you so choose. Simply click on them to turn them on.

When you have selected all the optional fields, click the **Next >** button to go to the next stage in the Database Wizard.

STEP 4: The next step is mostly cosmetic. You can choose how you would like your database to look based on several templates. Most corporate databases use the **Standard** option since it looks the most professional. Click the **Next >** button to go on.

*Most corporate databases use the **Standard** option since it looks the most professional.*

If the templates don't suit your needs, you can create a database from scratch.

STEP 5: Access now offers another style choice, this time for the printed report. If you want, you can have the display on the screen look plain, and then fancy it up for printing, or vice versa. Choose **Corporate** and click the **Next >** button again.

STEP 6: The next choice is what name you want to give the overall database records. Just type a name in the box — something like "My Company Expenses." Don't let this confuse you with the document's filename. Remember, we named the file way back in Step 2. This is just what you'll see when working with the database and on the reports.

STEP 7: Access also offers a choice on this screen to add a picture or company logo. To do this, check the box next to **"Yes, I'd like to include a picture."** Then the **Picture...** button lights up. Click it, then browse your way though the hard drive or network until you find the picture or logo you want. Select it and it will be placed in the corner of each and every database sheet. Click the **Next >** button yet again.

STEP 8: Click **Finish** to create your database. (*Figure 7.2*)

Depending on the speed of your system, it can take several seconds or even minutes to complete the database creation. Sit back and watch as Office XP keeps you apprised of its progress.

Going it alone

If the templates don't suit your needs, you can create a database from scratch. Simply run Access; from the **File** menu select **New**. When the Task Pane opens

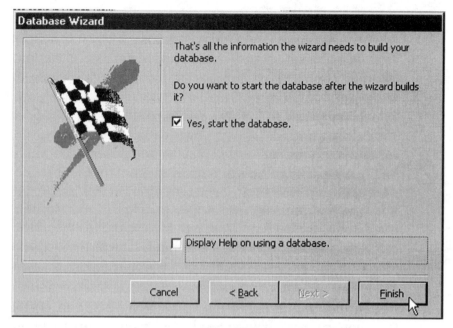

Figure 7.2

up on the right side, select **Blank Database**. Then name the file and click **Create**.

Three choices confront you when making a database from scratch. They are **Create table in Design View**, **Create table by using wizard**, and **Create table by entering data**.

The first option, **Design View**, is fairly easy. You are given three columns. Information must be entered into the first two to create a field.

The first column is the field name. The name of a field can be up to 64 characters long. Spaces count as characters. The second column lets you define what types of data users can enter into the field, be it numbers, dates, objects, or even hyperlinks. The optional third column is a more complete description.

Three choices confront you when making a database from scratch.

For example, you may want the first column to be "Name." In the second column, choose a type. Probably the entry showed up as "Text," so just leave that. In the last column, leave it blank or enter "Name of the Employee."

All fields can be changed once they are in the new table window.

The second option in creating a database uses the Table Wizard. It's probably the easiest. You are given a list of sample tables and fields. To add a field to a database, just click on the right arrow symbol to add it to the **Fields in my new table:** window, as seen in *Figure 7.3.*

This is the easiest way to make a table from scratch because almost every choice within a standard database is available. The more business-oriented fields like Social Security number and salary are found when the **Business** radio button is selected. The more personal fields like spouse's and children's names can be accessed when the **Personal** button is highlighted.

All fields can be changed once they are in the new table window by selecting them and pressing the **Rename Field...** button, so you are never locked into having a field like you are with the templates.

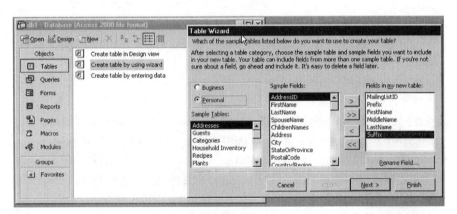

Figure 7.3

The final choice when making a new database is **Create by entering data**. This method is relatively simple as well, especially if you know exactly what you want in your database.

The columns represent records, and rows across the top represent fields. One record can have multiple fields. If you think this looks a lot like an Excel Spreadsheet, you are correct. Remember, for low-level inventory tracking, Excel is probably the best way to go.

If you think this looks a lot like an Excel Spreadsheet, you are correct.

Switchboard operator, can I help you?

Once you have set up all the fields you need, it is time to enter your data. When you initially create or load a database, you will be taken to the **Main Switchboard** screen.

This is a fancy name for the control panel that gives access to the different parts of Access. The name of the database you are working with will appear at the top of the switchboard.

The first option will be to **Enter/View** whatever type of database you are working with. In the case of a contact database the option will likely be called **Enter/View Contacts**. For our expenses database, the Switchboard's first entry is **Enter/View Expense Reports by Employees**. Clicking on this will bring you to the first record in the database, which will be blank if it is a new database or most likely a completed field on an existing one.

If you have a new database, then the opening field will be blank and ready to type on. Just select the first field and type in your data. Pressing **Tab** will allow you to

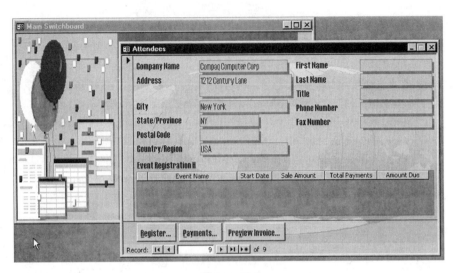

Figure 7.4

move between fields. Continue until the record is completely filled out or as complete as possible given the existing data you have.

When you are finished, click on the right-pointing triangle — the international symbol for Play ▶ — to advance to the next record. You'll probably find this near the bottom of the window next to the record number.

Continue until the record is completely filled out or as complete as possible given the existing data you have.

If you are working with a more complete database and want to enter new records, you can click on the right-pointing triangle button with a vertical line ▶| to advance to the last record. You can also click on the triangle with the asterisk ▶* to go right to the next blank record if you so choose (*Figure 7.4*).

When you are finished entering data into the records or ready for a break, just click on the **X** to close the window. Records are saved automatically, so there is no chance users will accidentally exit without saving work.

Operation: Data Retrieval

So now that you have some data entered into a simple database, what can you do with it? Well, the easiest thing is to perform simple searches to find the data you need.

Let's take a contact database as an example. Say you have a contact database with 450 records — our business contact database is about this large — and you want to find the person who works for Compaq. You can't remember the name, but you know the company (*Figure 7.5*).

From the Main Switchboard, click on the **Enter/View** button — just like you would to enter a record. The first record will appear. Then click on the field you want to search, in this case the company name.

Punch the **CTRL+F** buttons and the **Find and Replace** window will appear. Then you type in the search

Say you have a contact database with 450 records and you want to find the person who works for Compaq.

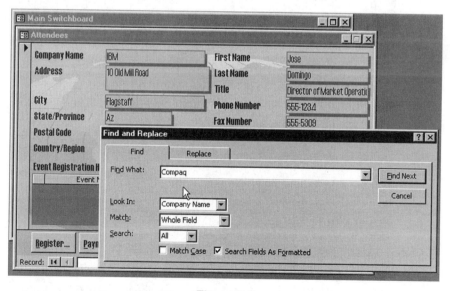

Figure 7.5

term, in this case "Compaq." When you click **Find Next** button, a search within that field will commence. The first record that contains a person working at Compaq will be displayed. If this is the right person, you can close the find window. Otherwise, keep clicking **Find Next** until the contact you want appears. In this way, you can have a simple yet powerful database at your fingertips.

You can customize what language you would like the spell check to use.

International Success

This time around, Microsoft has put a lot of new features into Access to help the international traveler. Since Access databases have the capability to be highly complex, it makes sense that in a corporate environment that this data might be shared internationally. That means different languages and even reading styles need to be supported.

New in Office XP

Choose language in spelling tab

When you click the **F7** button or choose **Spelling...** from the **Tools** menu, you will notice a new **Options...** button near the bottom of the window — that is, if Access finds a misspelled word. If not, you can get to this section by going to the **Tools** menu, selecting **Options...**, and then clicking on the **Spelling** tab.

Here you can customize what language you would like the spell check to use. It's a pretty diverse set of choices, too (*Figure 7.6*).

As a side note, Access now can automatically display multilingual text in tables and forms, so long as the

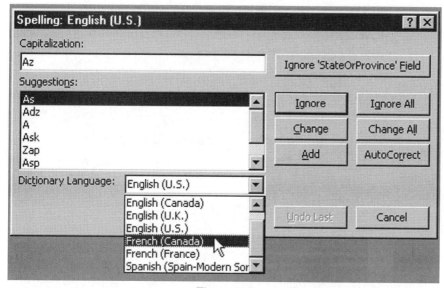

Figure 7.6

operating system that Office XP is running on has the proper support fonts installed.

There is also an **International** tab included in the **Options...** selection under the **Tools** menu. This allows users to change the direction of reading, from standard English left-to-right to right-to-left. This will ensure that if you are making a database in a language that reads other than left-to-right, that fields will be formatted properly.

New in Office XP

Change direction of text in fields

As another nod to the international Access user, the exporting feature has been improved when using international characters or mixed-language documents. As an example, if you have a report with multilingual text and you export it as a formatted rich-text document, when you open it in Word it will look correct.

The exporting feature has been improved when using international characters or mixed-language documents.

Advanced Shortcuts for Quick Fingers

When dealing with Access, there are probably a million ways to run a database. Entire books have been written on all the nuances of Access, but people who use these advanced features probably know what they are doing, and those who don't probably never will need to. So what we have done is to compose a list of new advanced features for the highest level of users.

It is no longer necessary to manually go and open up the Code window to get to all the advanced features.

If you don't use Access but would like to, the information we have given thus far will help you set up and maintain basic databases. The rest of this chapter however deals with new features inside the advanced modes of the database application. Expert users will find this helpful, since a lot of this is what we have been asking for over the past decade or so. But others might find it unnecessary reading. With that warning, let's continue.

New in Office XP

Shortcut!
Press F7 while in form or report design to enter the Code window

When users are inside either the form or report design window, it is no longer necessary to manually go and open up the Code window to get to all the advanced features. Simply pressing F7 does the trick now. Just remember that you have to be in the design window for this to work. Otherwise, you may end up activating the spell check.

When you are inside the property sheet design view, you can move the window focus back to the design surface. This is not new, but now you can do it without changing the control focus. Just hit **Shift-F7** and Access will do this for you automatically, meaning you should save a bunch of time when quickly creating field properties.

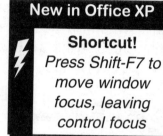

New in Office XP

Shortcut!
Press Shift-F7 to move window focus, leaving control focus

Another shortcut that can be found when in the design window is the ability to switch between the design window and the property sheet. Pressing **F4** accomplishes this task. Users found themselves constantly moving between these areas, so Microsoft made it easier.

New in Office XP

Shortcut!
Press F4 to toggle design to property sheet

There are times in extremely complicated documents where multiple views are possible. Instead of backing out and then back in to move between windows, Access now provides a set of keystrokes that will let users toggle among views as often as needed. These keys work with any query, table, page, view, stored procedure, or form.

When there is a multiple view possibility, you can use the **CTRL** and > key pressed together to move to the next view down. Pressing the **CTRL** and < key will move to the next view up. If you are at the last possible view, say when scrolling downward by pressing

New in Office XP

Shortcut!
Press CTRL-> and CTRL-< to move to new viewable menus

Another shortcut that can be found when in the design window is the ability to switch between the design window and the property sheet.

CTRL and >, then a subsequent push will loop back to the first menu in the series. Oddly enough, this also works when pressing the **CTRL** and **.** (period) keys to move down and the **CRTL** and **,** (comma) keys to move upward.

Besides being just a bigger magnifying glass, these new modes make it easier to examine the fine print on reports

New in Office XP

500% and 1000% zoom modes

Another new feature, while not really a shortcut, gives users more power to proof reports in zoom mode. Both a 500 percent zoom and a 1,000 percent zoom have been added to the option list. Besides being just a bigger magnifying glass, these new modes make it easier to examine the fine print on reports and to make sure that the tiniest of details do not ruin an otherwise flawless report.

New in Office XP

Shortcut!
Press F8 to toggle design view and field list

If you are in the form and report design view and you want to move to the field list, you can press **F8** to be instantly zapped over there. If you notice, most of these new shortcuts are taking place in the design view area. That is where users said the greatest need for shortcuts exists. Designing the forms is probably the most difficult aspect of Access. Once that is complete, entering the data is comparatively simple.

New in Office XP

Shortcut!
Pressing ENTER now adds selected field to form

A minor shortcut within the design view, but one that intuitively makes sense is the **Enter** key. Pressing **Enter** when you have selected a field in the field list form or report design view will add the selected field to the form or report design interface. And since most folks want to hit **Enter** when they are finished typing a line, this new feature is a long time in coming.

Some new hot keys are used more than once, with the resulting function depending on when they are activated. A great example of this is the **F8** key, which we previously ex-

New in Office XP

Shortcut!
Press F8 to shift focus to field list

plained will toggle between the design view and field list. However, when you are in the page-design view and press **F8**, the result is that the focus will shift to the field list.

When dealing with sub-sections, or with fields that are linked in some way to a sub-section, you can now toggle quickly between the subsec-tions. Pressing **CTRL+TAB** when you are in a report sec-

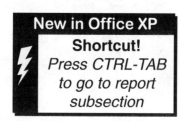

New in Office XP

Shortcut!
Press CTRL-TAB to go to report subsection

tion will take you to any existing subsections. Pressing **CTRL+TAB** again will take you back to the main report.

Some new hot keys are used more than once.

Really, really advanced stuff

The following new features are being pointed out for those power users among us who create Access objects in eXtensible Markup Language (XML) format, open da-tabases programmatically, or are running live or often updated databases off an SQL server. If you are not do-ing any of these things, you really won't need to know about these new features. If you are, you will wonder what took so long to get them implemented.

One of the best new features in Access is the ability to perform multiple redo and undo commands at the same time. In our tests, the multiple redo features worked in all of the following objects: ADP functions, forms, reports, macros, data access pages and modules, MDB tables, MDB queries, ADP views, and ADP stored procedures. However, be careful when creating undo stacks with the following objects: MDB tables, ADP stored procedures, functions, data access tables, and views. The reason is that the undo stack will be erased if you switch between views when using those special objects.

The undo stack will be erased if you switch between views when using those special objects.

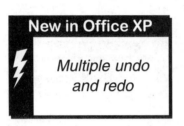

New in Office XP

Multiple undo and redo

New in Office XP

View object as PivotTables

If you need quick data analysis, it's good to know that now you can also view any MDB table or query, or any ADP table, view, stored procedure, or function as either a PivotChart or a PivotTable. You can find out more about PivotCharts and PivotTables in Chapter 4 on Excel.

And once you are finished with your PivotChart or PivotTable, you can now share the file with anyone, even users who don't have Access on their systems. There is an option under the save area that lets these tables and charts be saved as a data access page file. That way they can be viewed by

New in Office XP

View tables and charts in Internet Explorer 5 Web browser

anyone with Microsoft Internet Explorer 5 or better on his or her system.

True database programmers — the folks who get in there and actually code either the entire database or at least the behaviors of fields — will also find a couple of extra weapons in their arsenal.

New in Office XP

BrokenReference property added

One of the biggest time savers is that a new property, BrokenReference, has been added. When used, it will automatically detect any broken references without having to exit program mode and loop through each reference in the database.

If users of databases have a tendency to run into problems when setting up or printing a query, database administrators can now control both Access and printer properties right from the program mode. Use the Printer object and the Printer's collection to control this from the code level.

New in Office XP

Printer object and Printers collection to control output

Although some might consider this a security flaw, programmers no longer need to enter a password to get into a password-protected database. When entering a database programmatically, you can simply pass the database password phase. Database administrators, who face the task of keeping several data-

New in Office XP

Password avoidance

It will automatically detect any broken references without having to exit program mode and loop through each reference in the database.

bases running, but who might not be given a constantly changing code, will find this very helpful.

New in Office XP

Record exit trigger

Another code is the new OnRecordExit event, which can be helpful if records need to go through some special procedure as soon as they are completed or even viewed. This new code can also be accessed from the form level of a database if needed. It will trigger an event of your choosing every time a user leaves a current record, whether viewing it or modifying it in some way.

Access has been greatly improved from previous versions.

Most of the existing commands, such as dirty record triggers and the like, have not been changed and should perform exactly how users remember from previous versions. At least not all the world is changing for harrowed database programmers.

Final Thoughts

Access has been greatly improved from previous versions, and it shows, even when we are not talking about the standard or even the program interface. The best example is the new file format that databases are saved under when using Access with Office XP.

Basically, this new format makes it easier for computers to query large data sets, resulting in quicker responses and fewer or no lockups. We tested a few large databases that were created with Office 95, 97, or 2000, and compared the results with the same size file created with the Access that comes with Office XP.

In every case, the checkered flag went to the new Office XP Access. Sometimes the difference was only a few seconds, but at times the differences were quite noticeable. There is nothing worse than just sitting there waiting for the computer to give a response and not knowing how long the process will take. Those days should have died out with UNIVAC. With the new file format, mostly they do.

The point is that unless, for some reason, you are really in love with the format of a database, it's probably best to go ahead and convert the entire thing over to the new format. Even in the very limited times we were able to generate an error when running a conversion — and we are talking about several nights of cold pizza, caffeine-laden soda, and no sleep to even generate a single error — the new error tracker had us back on track within minutes.

If you are really afraid, you could always back up your existing database to a Zip disk or burn it to a recordable CD before making the conversion. In fact, we recommend you back up everything before conversion. That way, if something does go horribly wrong you can at least get things back to the way they were.

Finally, it is worth mentioning that Microsoft says the best operating system to have when running Access is Windows 2000. While this might seem like an attempt to sell more copies of the somewhat lackluster OS, it's interesting to note that Microsoft does not say this same thing about other products in the suite, which should run just fine under other versions of Windows.

Microsoft says the best operating system to have when running Access is Windows 2000.

We set up several machines with different operating systems and had them crunch identical databases.

Never ones to let an opportunity for heavy laboratory testing go to waste, we set up several machines with different operating systems and had them crunch identical databases.

Although no databases experienced any errors — even ones running old operating systems like Windows 95 — there were some performance advantages for the systems running Windows 2000. This amounted to about 10 to 15 seconds over Windows 95 and 98 computers, and about 3 to 5 seconds over those running NT. That's not a huge amount considering the size of the database we were using as a test, but if seconds count, you might consider Windows 2000 on your database computer.

We did not get into a lot of the client-server issues here because such a small percentage of the readers of this book would even care to read about it. But we did do some testing and found that all the things programmers expect should work with Access, only in most cases they do it a little bit faster than previous versions.

8

FrontPage

The Web is hotter than a booster rocket at liftoff, and Microsoft Office XP embraces the Web full throttle. It can create Web pages, publish to the Web, and use the Web in almost every way. All of the applications in Office XP do it.

So why have a tool like FrontPage?

FrontPage does more than create a Web page. It can be used to create and manage a whole site.

Go to almost any site on the Web, whether personal or business, and more than one page will pop up as links are clicked. So, several pages link to each other. Imagine if one page's name is changed. Does that mean you have to go into all those different pages and make the change?

Not with FrontPage managing your site. FrontPage keeps everything linked and notifies you when a link is broken.

FrontPage keeps everything linked and notifies you when a link is broken.

FrontPage also includes Web-exclusive features, such as hit counters to see how many people have visited a site or other dynamic elements.

For the new FrontPage 2002, Microsoft has incorporated some new features, like Office Web Server's SharePoint sites for teams. We'll cover more of this in Chapter 12.

So while all Office XP products can create a Web page, FrontPage helps manage a Web site.

A five-page personal site is very different from a 30-page corporate site.

Basic Concepts for Building a Web Site

What kind of Web site do you want to build? That helps determine how to construct a Web site. Sites require planning. A five-page personal site is very different from a 30-page corporate site, which is very different from a 16-page intranet.

Knowing your objective is the first step. Office XP provides some basic sites ready to go. When FrontPage 2002 opens, the common Task Pane for **New Page or Web** opens. To use the wizard to build a Web site, select **Web Site Templates...** from the **New From Template** listing. The **Web Site Templates** dialogue window appears (*Figure 8.1*). Choose one to build, perhaps the **Corporate Presence Web** and then click **OK**.

Now the wizard launches. For most of the different sites, a wizard appears to ask a series of questions to help complete the information on the site.

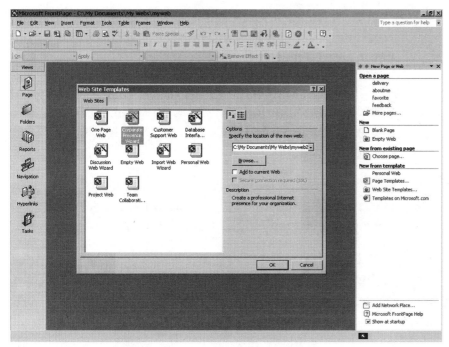

The next screen lists the pages that will be included in the site.

Figure 8.1

For example, after the introduction, click the **Next >** button. The next screen lists the pages that will be included in the site: **Home** (which is the first page seen and required), **What's New**, **Products/Services**, **Table of Contents**, **Feedback Form**, and **Search Form**. It's a small site for a small company, so let's uncheck **Table of Contents** and the **Search Form** pages. Click the **Next >** button.

The next screen allows you to choose what will be on the Home page or the first page people will see when visiting your site. The **Mission Statement** and **Contact Info** are already checked. Let's uncheck the **Mission Statement** and check the other two, the **Introduction** and **Company Profile**. Click the **Next >** button.

Themes give sites a coordinated look of colors, fonts, backgrounds, buttons, graphics, and more.

Now it asks about what should appear on the What's New page. Check **Press Releases** and uncheck **Web Changes**. Click **Next >**.

The wizard wants to know how many **Products** and **Services** are appropriate for your company. (We told you planning was important.) Let's choose two each. Click the **Next >** button.

You've got the idea. Just continue with the wizard by filling out information. Have some fun with it. Maybe make sure your company logo appears on every page.

Near the end of the wizard, you'll be asked to **Choose a Web Theme**. Themes give sites a coordinated look of colors, fonts, backgrounds, buttons, graphics, and more. Click the **Choose a Web Theme** button. Along the left side are the many selections to choose from. Click on some to see what the themes would appear like. Choose one. We selected Network Blitz.

Beneath the selections are options. The **Vivid colors** selection "boosts" the color palate, generally adding more hues. In some cases, it makes little difference. In others, it appears quite vivid. The changes appear in the mock-up.

Active graphics means some images and buttons will light or appear to move in some way when clicked or when a cursor hovers over it. You cannot see the impact of this selection until later.

Background picture, as you can see when you select it, will add a graphic to the background.

Apply using CSS refers to cascading style sheets. A CSS defines the fonts and colors of a Web site from a central document. A CSS can be handy if you're not going to vary from the way the pages of a site appear.

A couple of more steps and you're all done with your first Web site.

The Interface

After completing the wizard, the main display appears again, probably like *Figure 8.2*. In a way, FrontPage merges Word with a little bit of Outlook. Like the Outlook Shortcuts bar always on the left side of the screen, FrontPage offers its own **View** bar along the left side. After finishing the wizard, it defaults to **Tasks**. Just

*In a way,
FrontPage
merges Word
with a little bit
of Outlook.*

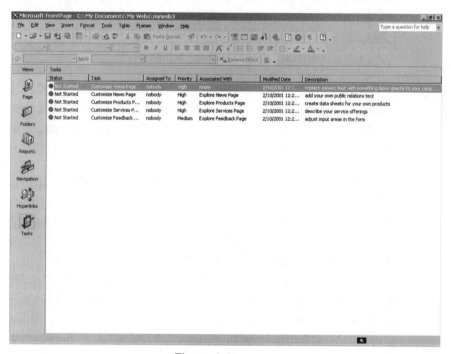

Figure 8.2

Microsoft Internet Explorer, Netscape Navigator, Opera, or any other browser will see the page and FrontPage can edit it.

like Outlook and its Tasks, FrontPage allows the Webmaster to assign tasks to complete as appropriate. Double-click the first one, which wants to customize the home page.

The Webmaster can assign tasks to others, rename it, and give it a priority of **High**, **Medium**, or **Low**. Go ahead and hit the **Start Task** button. The **View** switches to **Page** and then the index.htm page appears. Most Web sites default to the index.htm or index.html page. By the way, it doesn't matter whether the extension of a Web page is .htm or .html. Microsoft Internet Explorer, Netscape Navigator, Opera, or any other browser will see the page and FrontPage can edit it.

As you move around the page, notice how the cursor occasionally changes to an icon that looks a little like a hand holding a piece of paper. That's a **comment**. Double-click it and the comment will appear. Most of the comments require rewriting some text.

FrontPage allows graphical elements to be edited as well. See the page banner, as seen in *Figure 8.3*. We altered it. This is not a text element, but FrontPage generates it on the fly.

Now that we're in the Page view, notice the information along the bottom right. Ours says, "⧖ 10 seconds over 28.8." That's the approximate amount of time it would take for the page — as it currently appears — to load over a 28.8-kilobits-per-second dial-up connection. Most people nowadays have, at a minimum, a 56-kilobits-per-second dial-up connection. Just click on the area where the text appears and a drop-down menu shows with these choices: **14.4**, **28.8**, **56K**, **ISDN**, **T1**, and **T3**.

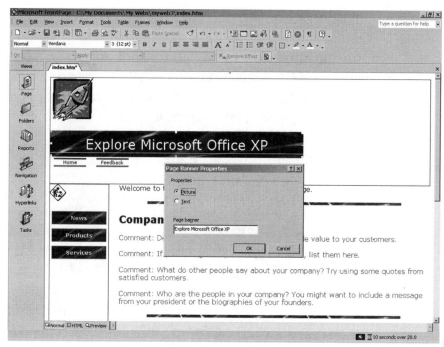

These are the different speeds that are available for connection to the Internet.

Figure 8.3

These are the different speeds that are available for connection to the Internet. Dial-up connections using analog telephone lines generally connect at **56K**, or 56 kilobits per second. That means for every second connected, 56,000 bits of data are transferred.

We don't want to get too math intensive, but 1,000 bits of data is about half of a page of text. Therefore, a 56K connection brings in about 28 pages of text that could be downloaded.

Of course, Web pages are made up of more than text. Images are heftier than just text, but at least you get the idea of the speed.

This foreign-appearing text is raw hypertext markup language, the code in which all Web pages are written.

ISDN stands for Integrated Services Digital Network. This was the predecessor to today's cable modems and digital subscriber line, or DSL, connections found in many homes. An ISDN connection runs at 128 kilobits per second. That's about 64 pages of text each second. A cable modem or DSL connection is about twice as fast as an **ISDN** connection. So if you're curious how long it might take someone to download a page from home using a DSL connection, the **ISDN** setting will come close.

T1 and **T3** deal with "trunk lines." These are among the large, fast connections across the country or into large companies. A T1 offers connections up to 1.544 megabits per second. A megabit is made up of about 1 million bits. This equates to a 772-page book being downloaded in a second. A T3 is even larger; it can send 43 megabits per second or about 28 copies of that 772-page book in a second. That's fast!

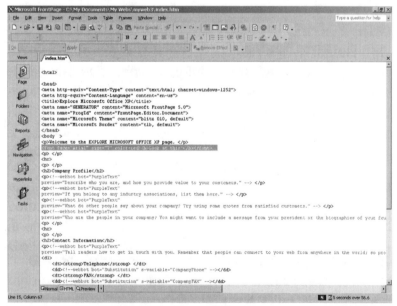

Figure 8.4

But for our purposes here, let's keep it simple and select **56K** from the list.

Along the bottom left, the **Normal** button is highlighted. It sits next to **HTML** and **Preview**. Go ahead and click **HTML**.

A lot of gobbledygook appears (*Figure 8.4*). This foreign-appearing text is raw hypertext markup language, the code in which all Web pages are written.

The basics of HTML are pretty simple. Information the Web browser needs to "draw" the page appears between the less than (<) and greater than (>) symbols. For example, look at this HTML code:

**<font face="Arial" size=7 color=red>
Look at this! </font>**

You can even type this in a new line while looking at the raw HTML. The < alerts a browser some code is about to follow. The **font** code lets the browser know we're going to be messing with the font information. So it's pretty obvious what follows:

face="Arial" .... change the font to Arial
size=7 .... make the text large
color=red .... make the text red

By the way, font size goes from 1 to 7 only in HTML. Browsers do not support larger text.

Size	Equivalent	Size	Equivalent
1	8 points	5	18 points
2	10 points	6	24 points
3	12 points	7	36 points
4	14 points		

The basics of HTML are pretty simple.

Of course, > lets the browser know we're done with font coding. The next code, **<center>**, tells the browser to begin centering whatever follows — in other words, turn centering "on." The next entry, ****, turns on bolding.

All of the text not encased in < and > appears on the screen of the browser. Therefore **Look at this!** appears centered in 36-point, red, bold Arial.

What follows turns off some codes:

**** .... turns off bold

</center> .... turns off centering

</font> .... turns off all of the font coding

*FrontPage offers a few more **Views** we've yet to see.*

Now click on **Preview**. This is how the page will appear in a browser. A little later in this chapter, we'll explore more about HTML. Go ahead and click on **Normal** again.

New in Office XP

Unified search-and-replace dialogue

One nice addition to FrontPage 2002 is a new, unified search-and-replace dialogue, similar to the one found in Word. Just press **CTRL+F** to access it. You can still press **CTRL+H** to access just the replace function.

Planning a Web Site

FrontPage offers a few more **Views** we've yet to see. Some of these **Views** provide good organization tools to understand a Web site. Click on the **Folders** icon. It prob-

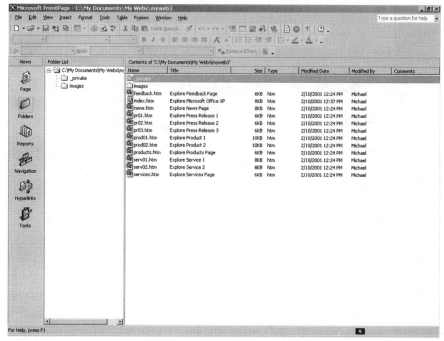

Larger, more complex sites will have several folders and folders within folders.

Figure 8.5

ably looks familiar because it's the appearance of how Web files are stored on a drive (*Figure 8.5*).

The Folders view helps where the content is stored. For example, all pictures by default get stored in the **images** folder. The **_private** folder is where feedback information is stored. For example, on our site is a Feedback page. You can see it if you double-click **feedback.htm**. When someone completes this form when this site is "live" on the Web, the information input will be put into a text file inside the **_private** folder.

Larger, more complex sites will have several folders and folders within folders to help organize the site better and make updating easier.

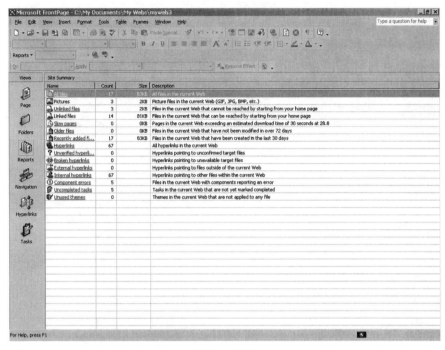

Figure 8.6

*When a site is live on the Web, the different options under **Usage** can provide feedback as to which pages are popular.*

Now, in the FrontPage **Views** bar, click **Reports**. The Reports tool can provide some good feedback on-site (*Figure 8.6*). It lists a variety of information, including the total number of files, pictures, and more.

While in the **Reports** view, FrontPage displays the **Reporting** toolbar. Click on the button with the word **Reports** in the toolbar. A dropdown menu appears. It contains most of the information in the main window except for the **Usage** ▶ selection. When a site is live on the Web, the different options under **Usage** can provide feedback as to which pages are popular and how often people visit.

New in Office XP

Usage Reports to see how many people visit your Web site

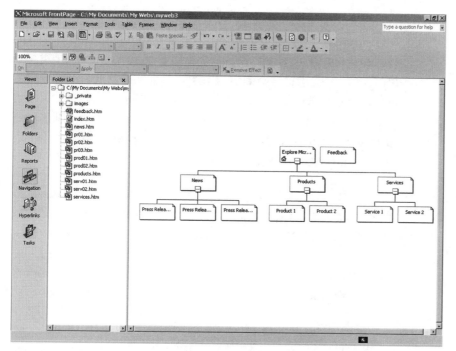

Figure 8.7

Now click on the Navigation icon in the FrontPage **Views** bar. A family tree-like diagram appears of the site (*Figure 8.7*). It shows how the site is logically organized. For example, from the Home page (that's the top one with a little house on it), a visitor can go to the News, Products, or Services pages. The Feedback page, which appears to sit off to itself, is a universal page accessible from all pages.

We can reorganize a little from here. See the main Services page. Let's move it under Products. Click with the left mouse button, hold it down, and drag the Services box until it is under Products. As you move it, a little line indicates the change. Release the left button and you're done.

The Feedback page is a universal page accessible from all pages.

241

Finally, let's look at the **Hyperlinks** selection under the FrontPage **Views** bar. When you first click on Hyperlinks, it looks a little like the **Folders** view. Just select a file in the **Folder List** window — for example, **News.htm**. FrontPage now dynamically draws all of the hyperlinks on page and where it will take you. Click on any **+** symbols for FrontPage to display where those hyperlinks go to (*Figure 8.7*).

Hyperlinks are the things you click in order to move to another page.

Hyperlinks are the things you click in order to move to another page. A graphic or text can be a hyperlink. Commonly, a text hyperlink appears as blue text with an underline, although it doesn't have to be blue and it doesn't have to be underlined. Whenever your cursor hovers over a live hyperlink on a Web page, it turns into a hand with a pointing finger.

FrontPage includes all of these tools to better understand how your site is organized and how visitors get between different pages of a site.

What is HTML?

Hypertext markup language continues to evolve. The World Wide Web Consortium, better known as the W3C, sets the standards for HTML. You can learn more at the consortium's Web site at www.w3.org.

New in Office XP

Support for HTML version 4.0

HTML can trace its roots to 1990. Today, HTML is a lot different. More than 500 member organizations and companies contribute to HTML as its evolution continues. Currently, HTML is at version 4.0.

Of course, a lot of other different technologies contribute to the Web experience. Some of these include Java, JavaScript, ActiveX, Macromedia Flash, Common Gateway Interface or CGI, and much, much more.

At its simplest, though, each .htm or .html file contains nothing more than text with the "code" hidden within the less than and greater than symbols.

FrontPage can keep you away from learning this code. Oh, it's very helpful to know HTML and the basics. It's like the hidden codes in a Word document. Understanding HTML allows a more granular control of any Web page's appearance.

Understanding HTML allows a more granular control of any Web page's appearance.

The Web itself is full of resources that teach basic HTML. Moreover, we found a good way to learn is by experimenting on your own. Build a page in FrontPage and then look at the code. Take some out. Alter it. You'll soon discover the different between a **<p>** and **
** (the first is a paragraph break like hitting **Enter** twice in a document, the other a line break like hitting **Enter** only once).

Also, in the *how'd-they-do-that?* department, simply check out the underlying code. Most Web browsers offer a way to see the raw HTML. In Microsoft Internet Explorer, you'll find under the **View** menu the selection **Source**.

But you need never do that with FrontPage. FrontPage keeps you safe from HTML.

Basic formatting

Let's begin with a blank page. Close the current Web site you're working on (from the **File** menu, select **Close**

If you decide to change the font, remember that the text will appear as you intend only if the viewer's PC also has that font installed.

Web). FrontPage remembers that you're currently working on that corporate-presence Web site. Therefore, if you were to create a new page, it keeps the Network Blitz or whatever theme you selected in mind. We don't want any formatting.

To clear it out, we need to create a new blank Web site. Under the **File** menu, select **New** ▸ and then **Page or Web....** The **New Page or Web** Task Pane will appear. Under the **New** listing, select **Empty Web**. When the window pops up, the **Empty Web** icon should be highlighted. Click the **OK** button.

With a blank page, let's first welcome people to the site. Type in a short message, something like "Welcome to My New Web Site." Text appears directly on the page. Notice how similar the toolbars across the top appear compared to Word (if you don't see any toolbars or want to verify which toolbars should be present, select **Standard** and **Formatting** from the drop-down menu after right clicking near the toolbars).

Within the limitations of the Web, the formatting of text is basically the same. Highlight the welcome message. Change the font, style, color, size, and more. If you decide to change the font, remember that the text will appear as you intend only if the viewer's PC also has that font installed. Generally, you're safe with Times New Roman, Arial, Tahoma, Courier New, Verdana, and a few others. However, fonts like Vivaldi, Eras, French Script, and Impact might not work since some people may not have installed those on their PCs.

Again, FrontPage in this mode is very much like Word. The formatting, while limited in some respects, still

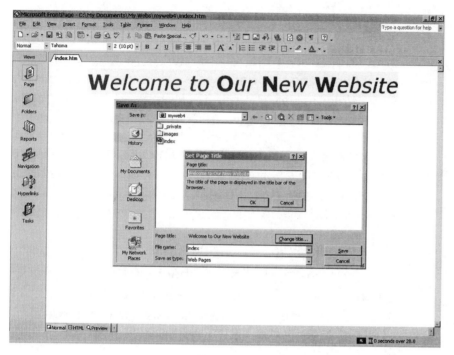

Figure 8.8

reacts and appears like Word. Just to check your progress, you can always click the **Preview** button to see what the page will look like in a Web browser.

Add some more text, perhaps a paragraph about yourself. Again, play around with some formatting until it looks the way you like.

Now, add a line of text like "Learn More About Me." We're going to make that a hyperlink a little later. But first we need to create that page about you. Save your current page as **index.htm**. When you select **Save** from the **File** menu, notice there's a small addition to the dialogue (*Figure 8.8*). The **Change title...** button next to the current title of the page is now there.

*When you select **Save** from the **File** menu, notice there's a small addition to the dialogue.*

The title of a Web page appears along the very top of the browser nearest the top left corner. For example, open a browser window and visit www.microsoft.com. Most likely, the title bar of the window says something like "Welcome to Microsoft's Homepage" followed by the name of your browser.

FrontPage picks up the first line of text automatically as the title.

FrontPage picks up the first line of text automatically as the title, but if you don't like it, click the **Change title...** button and type in your alternative.

Adding graphics of all sorts

Let's create that page about you. Close the welcome home page (select **Close** in the **File** menu). In the toolbar, click the **Create a new normal page** button, the first one on the **Standard** toolbar that looks like a blank page ▯.

Let's start out the page with a photo or graphic you like. You've probably got one on your hard drive somewhere. From the **Insert** menu, select **Picture** ▸ and then **From File...**.

Now locate that file on your hard drive. In Windows 98 and Windows Me, the folder **My Pictures** inside **My Documents** might contain some photos if you don't have any. Otherwise, the Windows directory in all versions of the operating system contains a bitmap image of some sort.

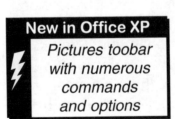

New in Office XP

Pictures toobar with numerous commands and options

Once you locate an image and insert it, FrontPage offers its **Pictures** toolbar (*Figure 8.9*).

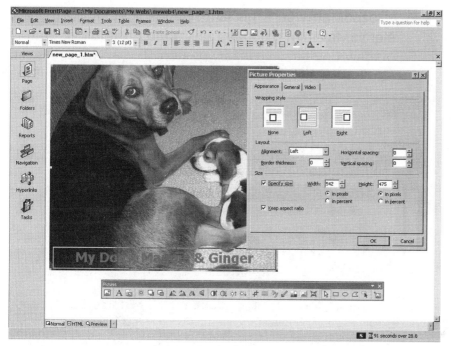

Figure 8.9

This is where FrontPage 2002 in Office XP begins to stand out above and beyond previous versions.

While FrontPage maintains some common elements, such as those found in the other Office XP applications, the Pictures toolbar includes some unique commands. For example, by hitting the **Text** button represented by the letter **A** on the toolbar, FrontPage will convert any image to the 256-color .GIF format and allow text to be inserted as part of the picture.

Right click the image and select **Picture Proper-ties...** from the drop-down menu. The Picture Properties dialogue offers several options including whether text wraps around the image, a border around the image, and absolute sizing. Sizing can be helpful for users' brows-

FrontPage will convert any image to the 256-color .GIF format and allow text to be inserted as part of the picture.

ers. If the connection is slow, the browser knows upfront that the image is so many pixels wide and tall.

Do not use sizing to make an image appear smaller than its native size. Let's say you have an image that's 500 by 700 pixels. If you use the size command here to reduce it to 250 by 350, the native picture is still larger; it just appears smaller on the Web page. So, the browser will download the larger image, taking twice as long to download since the larger the image, the larger the file size. Reduce the size from the main interface and select the **Resample** button in the **Pictures** toolbar. Then the image will be converted to the size you'd like and won't take as long to download.

FrontPage also incorporates a way to create so-called "hotspots" on images for hyperlinks.

From the **Pictures** toolbar, FrontPage also incorporates a way to create so-called "hotspots" on images for hyperlinks. These hotspots can be used in a variety of ways, just like all hyperlinks.

For example, let's say a page will be dedicated to each dog, Maggie and Ginger. Using the **Pologonal Hotspot** tool, an outline of each dog can be created. Draw an outline around the sections, somewhat like the connect-the-dots game. Clicking **Highlight Hotspots** will show the resulting hotspot areas (*Figure 8.10*).

Right click inside the hotspot to select **Picture Hotspot Properties...** to check where the hyperlink will go.

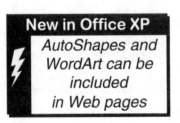

New in Office XP
AutoShapes and WordArt can be included in Web pages

FrontPage 2002 now supports AutoShapes and WordArt incorporated directly into Web

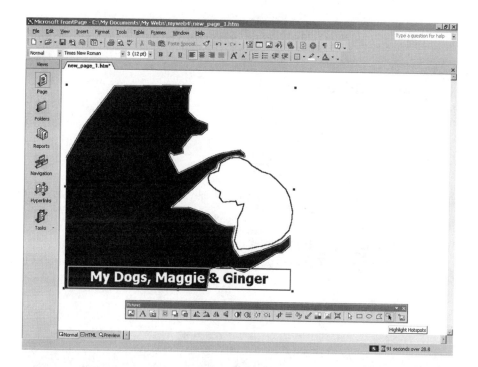

Figure 8.10

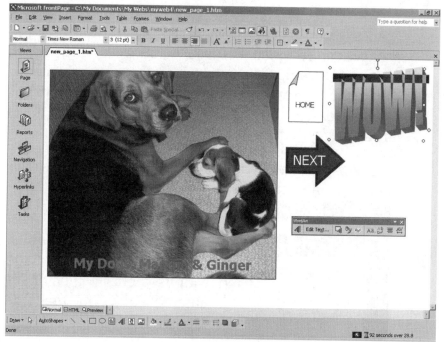

Figure 8.11

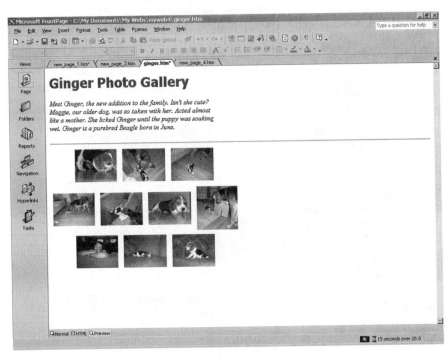

Figure 8.12

pages (*Figure 8.11*). Like other Office XP applications, these shapes can be added with ease. AutoShapes can be added from the **Drawing** toolbar. Add colored arrows, hearts, smiling faces, bursts, and all the figures in the AutoShapes tool. WordArt can be accessed directly from the **Insert** menu by selecting **Picture** ▸ then **WordArt...**.

A brand-new feature in FrontPage 2002 automatically generates a photo gallery, which is common to most personal Web sites. *Figure 8.12* shows one format of the gallery.

A brand-new feature in FrontPage 2002 automatically generates a photo gallery.

New in Office XP

Photo Gallery created automatically

FrontPage makes it easy, generating the page with its requisite thumbnails automatically. Two methods are available:

METHOD 1: Create a new page using the photo gallery template. From the **File** menu, select **New** ▸ then **Page or Web....** The **New Page or Web** Task Pane will open. From the **New from Template** section, select **Page Templates....** Select the **Photo Gallery** icon under the **General** tab and click the **OK** button.

METHOD 2: If you already have an existing page (or you create a blank page by clicking the **Create a new normal page** icon in the **Standard** toolbar), from the **Insert** menu select **Picture** ▸ then **New Photo Gallery....**

If you use Method 1, right click anywhere on the photo gallery and select **Photo Gallery Properties** from the drop-down menu. This pulls up the **Photo Gallery Properties** dialogue on the main **Pictures** tab — the same dialogue that appears when you use Method 2. If sample images appear, select them all and click the **Remove** button.

For your selection of images, click **Add** ▾. The options **Pictures from Files...** and **Pictures from Scanner or Camera...** appear. If the images aren't already in files on your hard drive, you can scan them in by using the second selection. But generally you'll be choosing **Pictures from Files....**

Select multiple files from the dialogue by pressing and holding the **CTRL** key while clicking on each file (*Figure 8.13*). By using the **Move Up** and **Move Down** buttons, the thumbnail placement will be changed. Add a **Caption** and **Description** to each photograph if you'd like.

The **Edit...** button allows some limited options like cropping or rotating an image.

If the images aren't already in files on your hard drive, you can scan them in by using the second selection.

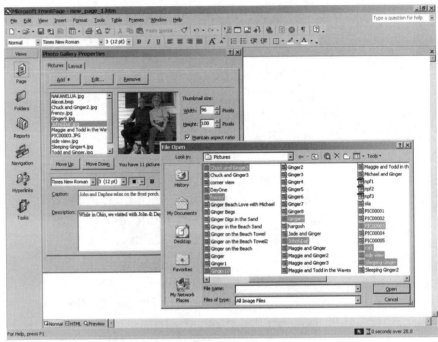

Figure 8.13

The **Layout** tab offers four different page designs, as seen in *Figures 8.14* through *8.17* on the following pages. Select the one you'd like and click **OK**.

Hyperlinks

Like a transporter on a starship, clicking a hyperlink takes you someplace else almost instantaneously. Of course, it depends on the bandwidth how quickly that transport happens.

As an experienced Internet user, you've probably clicked on thousands of hyperlinks. Images and text can be hyperlinks. Hyperlinks can open more windows or even send an e-mail.

Like a transporter on a starship, clicking a hyperlink takes you someplace else almost instantaneously.

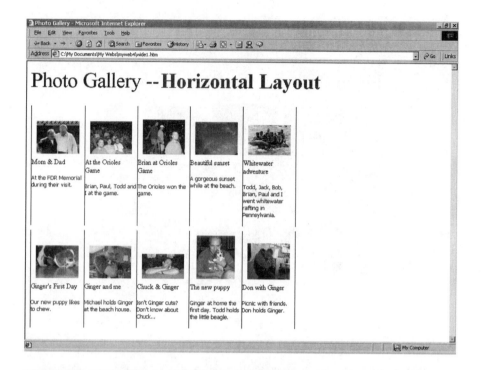

Figure 8.14

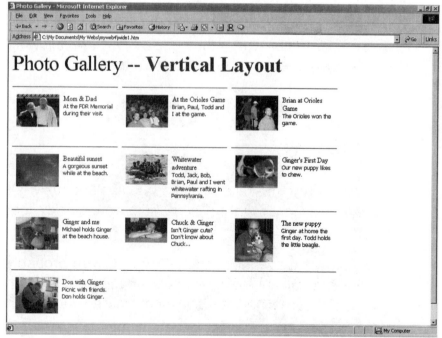

Figure 8.15

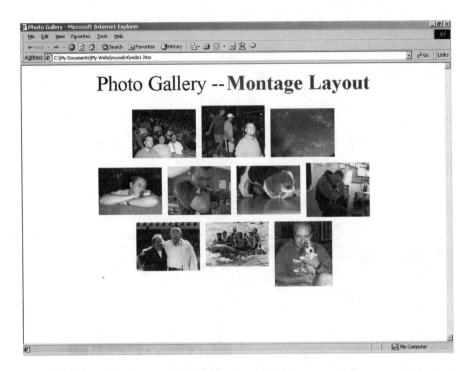

Figure 8.16

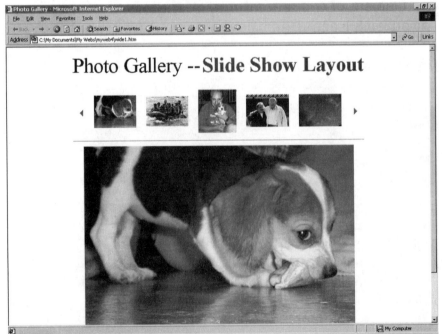

Figure 8.17

Early Web coders knew hyperlinks by its other name: anchor. This comes from the code used to create a hyperlink. For example, you may visit a page that includes a line like this:

Click <u>here</u> for additional information.

The code within the hypertext markup language, looks something like this:

Click <a href="http://www.microsoft.com/ office">here</a> for additional information.

Clicking on the word "here" will transport you to the Web site for Microsoft Office at <u>http:// www.microsoft.com/office</u>.

The **<a** introduction to the code indicates an anchor. The **href=** notifies the browser of a hypertext reference is about to follow. Of course, the Web site explains itself. The anchor end code is the familiar **</a>** indicating the end of the hypertext link.

While knowing and understanding a hyperlink helps, FrontPage makes it a bit easier. A hyperlink can be text or graphic. Go back to our index.htm page and highlight some text to create into a hyperlink. Now click the **Insert Hyperlink** icon in the **Standard** toolbar (it looks like a globe with a chain beneath it). Instead, from the **Insert** menu, select **Hyperlink...**. Or you can press **CTRL+K**.

New in Office XP

Comprehensive Insert Hyperlink dialogue with more options

With the **Insert Hyperlink** dialogue, a lot of choices pop up (*Figure 8.18*). The **Link to:** area on the left gives four choices:

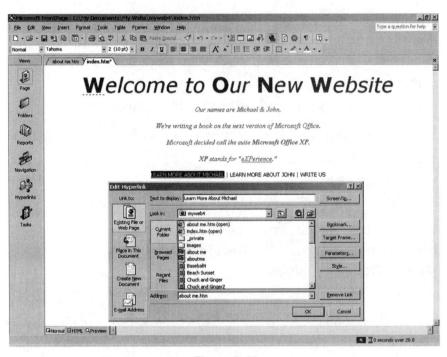

Figure 8.18

The file system area provides the recognizable entry for selecting files.

♦ **Ex̲isting File or Web Page**: Just as it sounds, you can link to another file or Web page. In *Figure 8.18* we selected the page named **about me.htm**.

♦ **Pl̲ace in This Document**: You can select a location within the document itself. These are known as bookmarks, taking you to different positions within that Web page. This can be handy for very long pages.

♦ **Create N̲ew Document**: Haven't created the Web page yet? Just select this one.

♦ **E-m̲ail Address**: When a visitor clicks this, the visitor's e-mail application launches and sets up an e-mail complete with your address. Be sure if you click this option that you type in your e-mail address.

The next section to the right is the **L̲ook in:** area, which provides for three selections:

♦ **Current Folder**: Lists all the current files within the folder. Since everything in this folder will be published to the Web, it's a safe bet to select an item here.

♦ **Browsed Pages**: This is to select pages on the Internet that you recently visited. You can also click the globe button in the top right to browse the Web and find the page for which you're looking.

♦ **Recent Files**: This lists recent files you've accessed including any Office document like a Word or Excel file.

The file system area provides the recognizable entry for selecting files. Beneath that in the field next to **Address:** you can enter a Web site address manually. For example, to link to Microsoft's Office site, you can just type in **www.microsoft.com/office**.

Along the right side are a series of additional buttons:

♦ **ScreenTip...**: When the cursor hovers over a button in the toolbar, notice how a small box appears giving the name of that button's command. A ScreenTip is basically the same for a hyperlink. If a cursor hovers over a hyperlink, the ScreenTip text appears. This is optional.

♦ **Bookmark**...: Unlike the Place in This Document entry, this links to bookmarks in other documents. For example, if you selected another Web page other than the one you're currently working on, you can use this button to link to a particular bookmark within that other page. Again, this is an option.

♦ **Target Frame...:** Have you noticed sometimes when you're on the Web and you click a hyperlink that a new browser window opens? This button can allow you to do that, too. If you click it and then select New Window, it will open up another browser window. This but-

Go back to our index.htm page and highlight some text to create into a hyperlink.

Other than just formatting the contents of a Web page, the page itself can be formatted.

ton is also used when a Web page uses frames, which we'll discuss a little later.

♦ **Parameters...:** This button is used for queries into databases.

♦ **Style...:** If the site uses cascading style sheets, a style can be applied to a hyperlink here.

♦ **Remove Link**: Does exactly that. If you have created a hyperlink, remove it by clicking this button. When you type text in FrontPage, if it sees a string it thinks should be a hyperlink, you can remove it here. For example, let's say you're writing about a local establishment called C@thy's C@fe. FrontPage will automatically think it's an e-mail address and will create a hyperlink. **Remove Link** gets rid of that.

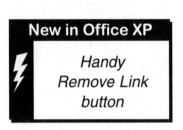

New in Office XP

Handy Remove Link button

By the way, if you decide to create a bookmark in any Web page, it's pretty easy. Like a hyperlink, highlight some text and from the **Insert** menu, select **Bookmark...**. A dotted purple line will appear while in Normal view, but you cannot see the dotted purple line when a page is viewed in a browser.

Overall page formatting

Other than just formatting the contents of a Web page, the page itself can be formatted. By right clicking anywhere on a page being edited, select **Page Properties...** from the drop-down menu. Most of the selections found in the **Page Properties** dialogue aren't needed. For example, on the **General** tab, you can select a background sound to play whenever that page is opened.

The **Background** tab is perhaps the most used. The background color of a page can be set here. Or, if preferred, a background image can be selected. Different hyperlink colors are also set here.

To Frame or Not to Frame

When building a Web page, sometimes it's convenient for portions of the page to remain the same while other portions change. Webmasters sometimes use frames to achieve this.

The browser's display area can be divided into sections. Each section, or frame, contains an individual page. For example, if you have three frames in the browser window, you're actually seeing three different files.

Experienced Webmasters are known to loathe frames. Primarily, frames can be bothersome because each visitor has a different-sized browser window. A frame's size can be absolute (so many pixels wide and/or tall) or relative (a percentage of the total window).

Moreover, it seems each browser supports frames in a different way. Frames can also be confusing to code properly.

In other words, it's probably a good idea to avoid frames altogether. If you decide to experiment, use the **Frames** menu.

Frames can be bothersome because each visitor has a different-sized browser window.

Creating Tables and Nesting Tables

Tables, on the other hand, should not be avoided. Just like the cells in Excel's spreadsheets, tables can be a very handy way to organize information.

In fact, creating a table in Word or formatting in Excel is very similar to FrontPage. However, FrontPage is a bit different since some minor formatting features in other applications cannot be applied to the Web.

FrontPage and browsers can handle something called "nesting tables."

To create one, from the **Table** menu select **Insert** ▸ and then **Table**. The number of rows and columns are needed in the **Insert Table** dialogue. Also pay attention to the **Specify width:** entry. The width can be absolute in the number of pixels wide or relative based on a percentage of the browser's width.

Keep in mind, like with frames, different people use different screen resolutions. So, a window's size can vary. Generally, most people use a resolution of 800 by 600 pixels, so you're safe with something slightly smaller.

Obviously, **Alignment:** determines whether the table is flushed toward the left, right, or in the center. **Border:** is the thickness of the line between the different cells and along the outside. You can also set this to 0 (zero) so there is no line.

Cell padding: puts some "padding" or space between the borders and the contents of the table's cell. **Cell spacing:** puts that space between the cells themselves.

FrontPage and browsers can handle something called "nesting tables." It means a table within the cell of another table (see *Figure 8.19* for an example).

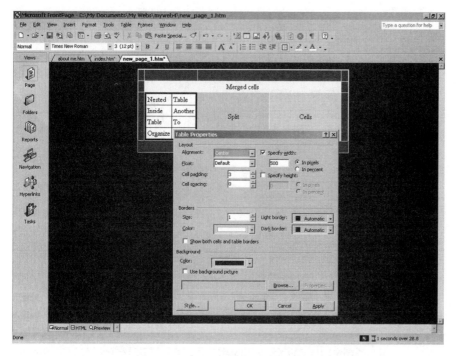

Figure 8.19

Individual cells can have different background colors and different types of contents. Two or more cells in the same row or the same column can also be fused together using the **Merge Cells...** command under the **Table** menu. A single cell can be split into two or more columns or rows using the **Split Cells...** command, also under **Table**.

FrontPage 2002 adds the **Split Table** command. If you had one table and wanted two tables, it required creating a new table. This command splits a single table between rows and not columns.

Individual cells can have different background colors and different types of contents.

New in Office XP

Split Table command to divide one table into two different tables

Somewhat like Excel, FrontPage can also fill in contents in rows or columns. Unlike Excel, FrontPage cannot fill based on some relative difference, such as progressing dates or increasing numbers. It will repeat the same content in the first cell. To access this feature, highlight the row or column to be filled and, from the **Table** menu, select **Fi̲ll** ▸ and then **Down** or **Right**, as appropriate.

New in Office XP

Automatically fill the contents of table cells down a column or right on a row

Experiment and check out the results in the **Preview** *view.*

Dynamic HTML and Java

Adding some dynamic effects in a Web site can be relatively easy using FrontPage's **DHTML Effects** toolbar. DHTML stands for Dynamic Hypertext Markup Language. These offer some basic effects similar to the effects in PowerPoint.

Right-click in the toolbar area and select **DHTML Effects** from the drop-down menu of the 12 different toolbars in FrontPage 2002.

For example, you can cause the HTML link colors to change whenever a mouse cursor hovers over that link. This is known as a "rollover" or "Mouse-over" effect.

Highlight a link and select when you want the effect to occur: Click, Double-Click, Mouse Over, or Page Load. As you build the effect, the appropriate selections appear in the next stage. Experiment and check out the results in the **Preview** view.

Java, of course, is another popular method for making a Web page stand out. Java is a completely different programming language and isn't something that can be accomplished with FrontPage 2002. However, let's say you found a Java applet you want to include.

To incorporate Java into a Web page, from the **In-sert** menu, select **Web Components...**. Several Microsoft-provided components are included here including a hit counter. In the left window, select **Advanced Controls**. In the right window, select **Java Applet** then click **Finish**. A straightforward dialogue pops up for inserting information on the location of the applet and the size of window it needs.

Java is another popular method for making a Web page stand out.

Saving Your Site

Microsoft continues to make saving and then publishing a Web site easier. When saving a site locally, FrontPage puts folders within the **My Webs** folder that's located within **My Documents**.

To publish to a live Web, it's as simple as clicking on the **File** menu and selecting **Publish Web...**. If you've got a Web site already established, just enter the normal directory to access it. You may need to enter a username and password. You can even browse to the site like it was your own computer's hard drive.

Of course, this is all just the beginning to creating a jazzy, cool, crazy, funky, wild, fun, professional, expressive Web site. Simple is better, but have some fun with all the options Microsoft FrontPage 2002 offers.

9 The Rest of the Family

Believe it or not, even if you buy the most advanced version of Office XP, you won't get every application in the Office family. Some applications need to be purchased separately.

Think of Publisher, Visio, MapPoint, PhotoDraw, and Project as allied worlds. Not quite part of the main federation of Office applications, each maintains the look and feel like one of the main lineup.

Separate but equal?

There are several reasons why these five applications are not packed into any version of Office XP. In one sense, separate applications may seem like a throwback to the days described in the first chapter, when each application had to be purchased separately. With Office XP, it is a bit more than that.

In the case of Visio, Microsoft only recently acquired it. Perhaps Microsoft did not have enough time to inte-

Think of Publisher, Visio, MapPoint, PhotoDraw, and Project as allied worlds.

grate it into the suite. More likely, Visio is basically a business process and network diagramming tool. The percentage of Office XP users that would make use of the tools is comparatively small. Office users probably would not be willing to pay extra money since only a specific group would probably make use of it. It's a fairly high-end program.

The timely nature of MapPoint may require a new version annually.

With MapPoint, the opposite is probably true. The application can give directions and street maps to and from almost any address in the United States, Canada, Mexico, or Europe, depending on the version. This application probably appeals to a wide audience since everyone travels to some extent. So it probably makes more economic sense to sell it as a separate program.

Finally, hundreds of engineers and programmers work on these applications. Getting each ready for release at the same time can be a complex scheduling challenge. Many applications are on a different development schedule. The timely nature of MapPoint may require a new version annually. A diagramming tool like Visio doesn't need a new version often. We did not see new versions of all of these applications, but new versions may be forthcoming in the months following the release of Office XP. Much of what we write about here focused on current versions available when Office XP launched.

Regardless of the reason for the separation, these applications feel just like any other Office application. Our goal with this chapter is not to go into great detail about every aspect of these applications. Instead, we will tell you what they do, and what you can do with them. This way, you can decide if you would like to pay the extra money to bring these applications into your personal universe of Office. Don't need them? Don't waste your time — and money — if they won't suit your needs.

Publisher

Publisher is a revved-up version of Word with more publishing than processing. We used to call these sorts of applications "desktop publishing tools." Publisher should be considered a midrange application in that it offers more functionality than Word, but less than advanced desktop publishing applications like Adobe PageMaker or QuarkXPress. Publisher is specially targeted for the home, home office, or small business user.

Although you won't get the same advanced functions you would with an application such as Quark, what you do get is simplicity and wizards that can walk you though projects. Publisher comes with two higher-end versions of Office XP, though most people won't get it right away, since it is not in the standard pack.

The interface in Publisher is very similar to Word. It seems every detail has been designed to mimic the functions in Word. For example, when you are making a bulleted list and press **Enter**, the application automatically adds another bullet to your document. This is a great feature since users are likely to know Word and therefore greatly reduces the learning curve for Publisher.

The wizard offers 20 different templates for Publisher. Some of them can be seen in *Figure 9-1*. Although this is a large variety and can do everything from make a sales

Publisher is a revved-up version of Word with more publishing than processing.

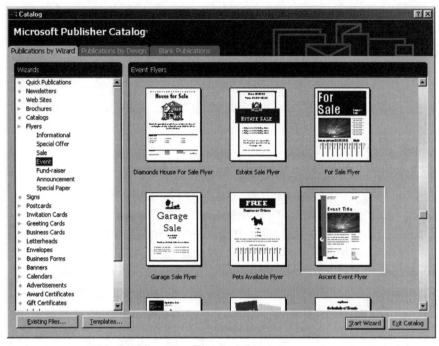

Figure 9.1

brochure to a company Web page, the problem is that your brochure could end up looking like someone else's. This problem can be eliminated in the wizard, however, by the use of an additional tool. When you make your document, a wand appears over different elements of the publication. Clicking on the wand can change different elements, such as page borders, or adjust overall colors. This can make your publication distinctive and look less like a standard template.

You can separate your brochure or publication into four colors for professional printing known as CMYK conversion.

There are a few additional features that are worth noting. You can separate your brochure or publication into four colors for professional printing known as CMYK conversion. For printed documents, this is a necessity if you need accuracy from your professional printing house. Any full-color document prints using four colors: cyan (a sky blue), magenta (a pinkish red), yellow, and black. These four colors, known as CMYK, combine to make all the colors in the rainbow.

So, Publisher is a great choice for offices that occasionally need to do desktop publishing, need more functionality than the standard Office XP application can offer, and don't have the time or desire for workers to get a doctorate in using the more complicated publishing applications. It's also pretty inexpensive on its own at around $120.

Visio

Visio is probably the most complex of the nonstandard Office XP family. With Visio, you can create simple images such as the organizational chart shown in *Figure 9.2*.

Visio is probably the most complex of the nonstandard Office XP family.

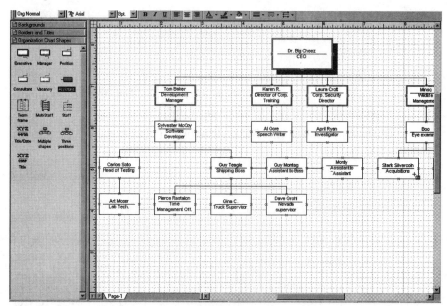

Figure 9.2

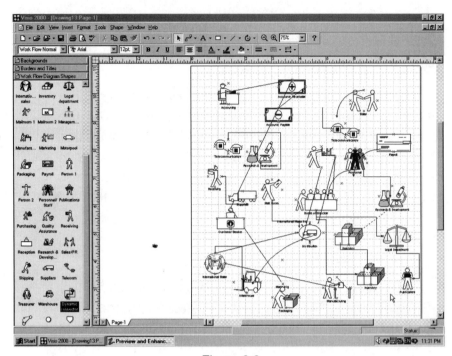

Figure 9.3

*At its core, Visio
is a visual
database.*

Visio can also do amazingly complex diagrams on everything from network infrastructure to the layout of who sits where within a new office.

At its core, Visio is a visual database. One look at *Figure 9.3* and you can see just how monstrous a Visio document can become without making the application even blink.

Microsoft offers four versions of Visio: Standard, Technical, Professional, and Enterprise.

The Standard edition offers more robust diagramming tools such as those found in other Office XP applications, especially PowerPoint. But Visio can create organizational or flowchart diagrams along with other types like timelines or even a project calendar. Standard also offers the shapes for office layouts with desks, chairs, conference-room tables, computers, and more.

The Technical edition takes charting one step further with network shapes for specific diagrams. Everything in the Standard edition is included, but Technical is specifically targeted to mechanical and electrical engineers, architecture, and building maintenance. Some basic computer-aided design, or CAD, is included as well.

The Professional version contains all of Standard plus greater focus on network and information technology design. Draw a local-area network or wide-area network. Additional business tools also are included with Professional.

The Enterprise edition takes everything a step further. Enterprise can handle detailed wiring closet diagrams.

One of the best features in Visio Enterprise is the AutoDiscovery tool. What you do is set up your copy of Visio on your network somewhere and then run the AutoDiscovery. Visio will send out feelers all along the network, riding on the simple network management protocol backbone. It will discover every device on the network and make note of how those devices connect to your switches, routers, hubs, and servers.

It then takes this information and displays it visually for you. Interestingly enough, each device will be represented by an accurate silhouette in the diagram. The Visio team does an excellent job of keeping the representations of the different devices up to date.

Visio will send out feelers all along the network, riding on the simple network management protocol backbone.

If you want, you also have the option of creating a visual database from scratch.

A server from Compaq Computer Corp. is going to look exactly like the actual device, right down to the grill venting, and in turn is going to look different from a server from Dell Computer Corp. This is helpful, not only for accuracy purposes, but also gives technicians a visual record of the network. If they are told to fix the Compaq server, they can check the Visio diagram and see, at a glance, where the server is located.

Another use for Visio that is less published is its ability to take existing relational databases and display them visually. You simply tell it where your database is, and it will go to work reverse engineering it into a visual structure. We tested this function with a variety of databases including Access, Microsoft Visual FoxPro, IBM's DB2, Oracle, Sybase SQL, and Corel's Paradox. It worked fine each time and crunched the numbers quickly, even on an aging Pentium II-based computer. You are in complete control of the transformation, as Visio's wizard lets you specify which fields should be visually represented. You also can modify how the information is displayed.

If you want, you also have the option of creating a visual database from scratch, though this is a much more complex process. It's probably best to do your database in Microsoft Access first and then have Visio translate it for you once you are complete.

Your visual database remains up to date as well. When you initially tell it to translate the original database — which remains intact and unmodified because Visio basically makes a copy — you can also set the application to take a snapshot of the original data. This can be done periodically or manually from within Visio.

What it does is to check the original database and compare it to the Visio visual version. If there are differences, you can sync either of the databases to come in line with the other.

So, office managers, network administrators, or database professionals could love Visio. The price tag is a bit steep. Expect to pay from $199 to $999 for Visio, depending on the edition. If you need the functions it has, it is about the best way to complete your task.

MapPoint

If you happen to be at one location and need to get to another, MapPoint can be your guide. It's actually a lot more advanced than a simple mapping program, and can actually function as a scaled-down geographical information system. MapPoint just joined Office as a business mapping application that incorporates a variety of analysis tools. MapPoint integrates well with Access, Excel, and Outlook in particular.

Experienced Office users find MapPoint generally easy to understand the menus and commands, knowing almost instinctively how to navigate. MapPoint includes a Data Mapping Wizard under the **Format** menu that's extremely powerful. Download or create data in almost any format and the Data Mapping Wizard can interpret it and plot it on a map. Download some state-by-state data from the Census Bureau, import it into an Excel spreadsheet, and MapPoint will show how to map the data. MapPoint can even handle county-by-county data.

MapPoint just joined Office as a business mapping application that incorporates a variety of analysis tools.

Simply by typing in a street address, MapPoint displays a map of the location.

The main use for MapPoint is probably going to be finding your way from place to place. Simply by typing in a street address, MapPoint displays a map of the location. The application is pretty accurate and, even in worst-case scenarios, is mostly only a couple of hundred feet off, mostly on residential streets. When you consider it maps the entire United States (another version maps Europe), then the amount of data is highly impressive.

If you tell the application where you are and where you want to go, it will map out the trip route for you. It will tell you the estimated miles, the estimated cost, and how long it will take to get there. You can then modify your route by say, taking back roads, and see how this will change your time of arrival.

Moreover, you can flag a few extra destinations to stop at on the way from here to there. MapPoint can tap into Microsoft's Expedia Web site and notify you about areas with construction or other possible delays.

For more local travel, let's say you might be on the road and need to make several stops. Fuel prices can be steep. MapPoint can organize the stops in an order that saves gas.

Maps from the application can be downloaded to PocketPC handhelds. You can draw on maps, highlight routes, print, and do pretty much everything else through the interface. And the maps are highly detailed, as shown in *Figure 9.4*.

MapPoint could eliminate the old standard business sales meeting where people tack pins to the wall to show where target markets are. It's also a lot simpler to use than a full-blown geographic information system application, so

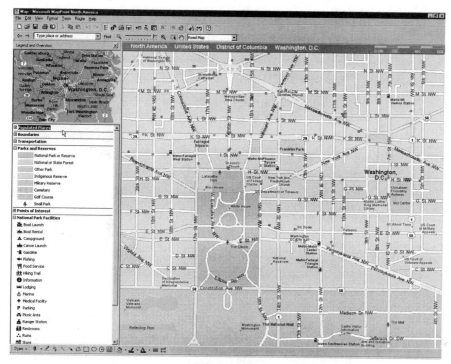

Figure 9.4

it should appeal to midsized and smaller business that can't afford to hire specialized employees to manage mapping data. It's also a lot cheaper at around $250.

In addition to the standard maps, you can also pull data on airports, universities, subways, public transportation areas, and even marketing demographics about different locations.

PhotoDraw

Microsoft included its newest edition to the Office suite in Office 2000, but PhotoDraw is dropped from Office XP. When PhotoDraw was first released, its sluggish

Microsoft included its newest edition to the Office suite in Office 2000, but PhotoDraw is dropped from Office XP.

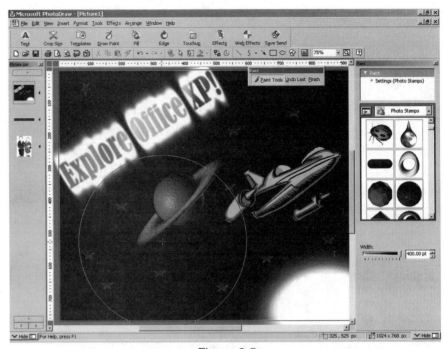

Figure 9.5

PhotoDraw is not your ordinary illustration or photo manipulation application.

---

response and quirky interface didn't stack up with the other Office applications.

Not so long after Office 2000, Microsoft revamped PhotoDraw completely and released PhotoDraw 2000 Version 2. Much improved and substantially faster, the new PhotoDraw could be just out of synch with the other applications. Or perhaps the original lukewarm reception is keeping engineers busy on yet another revamp.

PhotoDraw is not your ordinary illustration or photo manipulation application. It tries to do both. Like Publisher, PhotoDraw can be considered a low-end version of Adobe Photoshop merged with Adobe Illustrator.

Normally, an illustration application is separate from the photo manipulation tool, but PhotoDraw does both.

PhotoDraw comes with some powerful features to create anything from little buttons for a Web site to backgrounds for a PowerPoint presentation to all kinds of graphics in between (*Figure 9.5*).

Microsoft continues in the tradition of the wizard. Creating a relatively sophisticated image just takes a few steps. PhotoDraw comes on three CD-ROMs, two of which contain clip art and other image tools that make it so easy. Want a Web site logo or greeting card? Just select the appropriate wizard. PhotoDraw steps through it, changing colors, design, and fonts. Or create an image from scratch.

PhotoDraw is relatively easy, but not superior to other exclusive illustration tools. However, if you're creating a lot of PowerPoint presentations and need some special backgrounds, PhotoDraw might be right for you. And PhotoDraw isn't expensive — only about $110.

> *PhotoDraw is relatively easy, but not superior to other exclusive illustration tools.*

Project

Project is another tool in the Office XP family that was omitted from the main product lineup because of the limited amount of Office users that would need it. Project is very good for what it does — coordinating projects including reports, assignment of work, status updates, and time management. It is designed to work on a large scale with many people accessing the information from potentially different platforms. *Figure 9.6* shows Project being used to plan the construction of a house.

The newest version of Project keeps the project data on a central server. Information is then given to

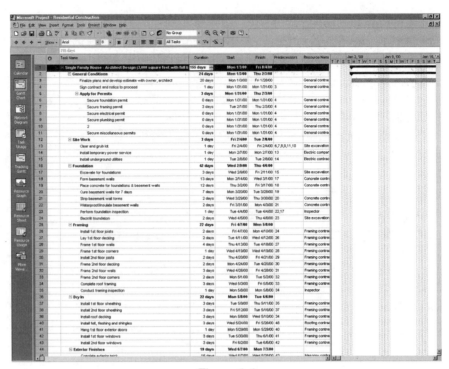

Figure 9.6

Low-end users only see data on their part of the project and may not even know what the larger goals are.

---

users — the folks working on the project — by Web-based updates through the Project Central component of the application. The person in charge of the project can control who sees what information by use of a dynamic editor. Low-end users only see data on their part of the project and may not even know what the larger goals are.

Say you are building a shuttle. The person assigned to making bolts would have information on what is needed and a timeframe, but need not know that their ultimate destination is a shuttle. The same goes for the computer system makers and the booster fuel team.

Users can be given rights to update the appropriate portions of the project as well, so the bolt maker can update the status of how many are created, their tolerance levels, and estimated completion times. A user who might depend on having the bolts, say the person building the shuttle, could see the bolt maker's plans, but would not ideally be able to change that data.

A further advantage to this setup is that the clients do not have to load a bit (or a byte) on their PCs. Moreover, since it is all Web based, everyone can see the data whether on a PC, Mac, or Unix platform.

Project can also provide cost and work estimates as the project continues and users update the status. Helpful elements such as risk assessment have also been thrown into the mix for good measure in addition to the standard schedule tools.

So, if you are the project manager for a large organization or work in an organization that has a less than stellar project management reputation, you might want to look into getting Project. It costs around $450. Smaller organizations can probably make do with the scheduling functions included in Outlook, though these will be quickly stretched to the limit when dealing with many users dispersed over great distances.

> *Smaller organizations can probably make do with the scheduling functions included in Outlook*

Pay to Play?

Whether or not you decide to invest in these nonstandard members of the Office XP family really depends on your job and needs. Probably 95 percent of the Office XP users can get by with the standard set of applications. But it is nice to know that if you need a little extra muscle, you can get it and not have to relearn a whole new interface, since most resemble the basic package.

There is no sense in visiting a new planet and having to learn about the culture and the language if you can go to one that speaks the way you do. Office XP users will feel at home using any of the nonstandard family of applications.

Office XP users will feel at home using any of the nonstandard family of applications.

10 Strictly Speaking with Office

The dream of being able to speak directly to your computer and have it understand what you mean is starting to become a reality. Once confined to science-fiction movies and television shows, we have advanced to the point where our faithful computers can recognize what we say in a dictation manner and display our words on the screen. In rarer cases, voice can be tied to a command or set of commands, so that a spoken command made to your computer can make it change directories, turn itself off, save or load applications, or do all the above in a predetermined order.

One of the coolest things about Office XP is that it has its own speech engine built right into the suite. This powerful application was created by Microsoft to work with Office XP. The elimination of a third-party speech application has resulted in higher accuracy rates. In our tests, the speech engine that ships with Office XP is somewhere between 96 and 99 percent accurate, depending on the

Once confined to science-fiction movies and television shows, we have advanced to the point where our faithful computers can recognize what we say.

sound equipment used and the acoustic conditions (i.e., background noise) around the computer. A computer sitting on a noisy factory floor is going to have a harder time detecting speech than one inside a quiet office.

> *When two people are having a conversation, a lot of the burden of comprehension is placed on the listener.*

General Dynamics of Speech

When a student begins to study how people communicate, one of the first things learned is how terrible humans really are at translating their thoughts into language. When two people are having a conversation, a lot of the burden of comprehension is placed on the listener. It might be surprising to learn how little of the actual message is carried by the language itself and how much is carried by body movements, rules of context, and even previous communications with the same person.

Besides the fact that a speaker might pronounce a word incorrectly or mumble a sentence, there is the problem of similar-sounding words that have different meanings — such is the problem with homophones.

Take for example this funny statement: "'I see,' said the blind carpenter as he picked up his hammer and saw."

What does that sentence mean? Did the blind carpenter suddenly gain sight or was he just gathering his tools? Granted, the statement is designed to fool you for a laugh, but even simpler sentences can have different meanings.

"I am seeing red," for example, means something completely different when you are picking out colors for your new dining room than when being reprimanded by your boss.

The computer has no way of knowing exactly what a speaker means, because it does not know the speaker's mood and can't read body language. In addition, the computer may interpret the former crimson-minded statement as, "I see read." Sure that sentence makes no sense, but when you say it out loud — the only thing the computer has to go on — it sounds the same.

For these reasons, speech recognition on the computer is a lot different than just spewing words onto a tape recorder. It's much more difficult than just recording sound. The computer is adding another step into the process. It is tasked with taking your words, which are translated thoughts, and turning them from the spoken word into written language.

The authors of this book have reviewed perhaps 10 different forms of speech-recognition engines over the past decade. We can vouch for the fact that the Microsoft speech engine is quite good. In fact, a couple of years ago, we visited Microsoft's Redmond, Washington, campus. While there, an engineer made an off-handed comment about Microsoft's speech-recognition engine. We pounced on that statement, asking more about it. The engineer told us that the engine wasn't ready. While it worked like other speech-recognition applications available, Microsoft wanted a more accurate and reliable engine before launching it. Looks like Microsoft got it.

Context rules are programmed into the engine. This helps the computer choose the correct form of a word

Microsoft wanted a more accurate and reliable engine before launching it.

based on previous use within a document and other words within the sentence. In fact, it's pretty difficult to trick once Office XP gets to know your patterns.

When we said Office XP is good once it gets to know you, we meant it.

Listen to what I say

Unless you specifically turned the speech-recognition engine on during the installation process back in Chapter 2, you will probably not have it ready to go by default. The first thing you will need to do is activate the engine. To do this, you will need to have your installation CD-ROM handy. You will also need between 15 and 30 minutes of free time.

The free time is not for the installation process itself; instead, it will be used to train the speech engine to your voice patterns. When we said Office XP is good once it gets to know you, we meant it. As with most speech-recognition engines, the more time you spend training as to what you sound like, the higher the accuracy rate and the more productive you will be when using speech instead of typing. But we will get into that in a minute. First, let's activate the engine.

Since you will likely be using Word for speech recognition, go ahead and load the application. Once you have your blank document up, go to the **Tools** menu and select **Speech** as shown in *Figure 10.1*. If the speech engine is not installed, you will get a warning message saying that Word cannot load speech files and asking if you would like to install them. Choose **Yes**.

Once chosen, you will be asked to insert the Office CD-ROM with the speech files. Installation is very quick and you should then move into the training program.

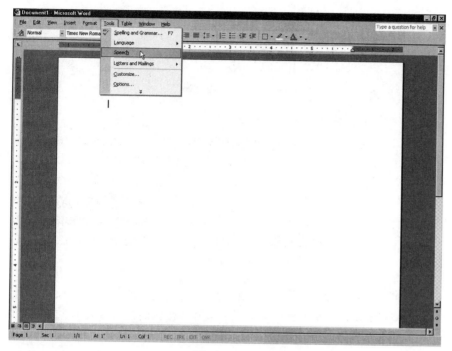

Figure 10.1

Let us make this one point perfectly clear. No speech engine — and Office XP is no exception — will work well without training. And the more you train your system, the higher your results will become. When you are finished with the first session, the accuracy rate will be somewhere around 85 percent for most people. Further training — even running the same sessions over again counts — will boost accuracy. Anyone should be able to boost his or her accuracy up to 95 percent, and many will see benefits close to perfect after extensive training.

Training is a simple process. The first thing you will be asked to do is to adjust your microphone, as shown in *Figure 10.2*. For best results, invest $20 or so in a high-quality headset microphone. Standard computer micro-

When you are finished with the first session, the accuracy rate will be somewhere around 85 percent for most people.

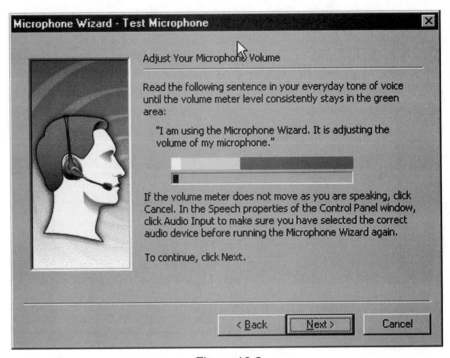

Figure 10.2

phones will work fine, but the headset microphones often come with noise-dampening technology that is ideal for speech. In addition, most come with at least one earpiece — if not two — that offers audio feedback.

The best way to adjust your microphone is to put the sound receptor a thumb's width away from your mouth. The microphone should be sitting slightly to the side of your mouth as well. This will enable it to pick up all sounds, but not breathing noises that can be accidentally interpreted as words by the engine. Also, if your microphone has an indicator showing which way the microphone inside the protective covering faces, make sure it has not been twisted around. Ideally, your microphone will be positioned as shown in *Figure 10.3*.

Figure 10.3

Say it like you mean it

Once the microphone is adjusted, your first training session will begin. Training consists of reading sentences displayed on the screen. As you say each word, the computer will highlight it to show that the word was recognized. Depending on the speed of your processor, the highlight process might lag a few words behind.

When you are training, say the words as you normally would without sounding mechanical. Speak at your normal pace, though concentrating on each word is helpful. This is the time that the computer is learning how you sound, so don't sound different than you normally would.

Also, don't worry if you pronounce a word incorrectly and the computer recognizes it anyway. The

When you are training, say the words as you normally would without sounding mechanical.

speech database is highly redundant, so one mistake won't ruin your profile.

If you keep reading and the computer does not follow you, it has become stuck on a word. Remember that the system might be a few words behind, but if you suspect the computer is too far behind, pause for a second and see if the system catches up. If it doesn't, then the first nonhighlighted word is the troublemaker. Say that word clearly until the system recognizes it and then continue reading from that point onward. You probably spoke too quickly, had noise interference, or just said something so differently the first time around that the engine got confused. If repeated attempts do not resolve the problem, you can press the **Skip Word** button. However, if the engine keeps missing words, try readjusting your microphone or reducing the ambient room noise if possible.

When you have finished the initial training session, you will be given the opportunity to train again.

When you have finished the initial training session, you will be given the opportunity to train again. This is highly advisable since initial accuracy rates are generally pretty low. Choose something interesting from the list of available topics, as shown in *Figure 10.4*. You have a pretty diverse list of subjects to read, from speeches made by Microsoft Chairman Bill Gates to excerpts from *The Wizard of Oz*.

These new sessions will be conducted exactly like the original one, with you reading words and the computer learning what you sound like and how you speak. When you are finished with a section, your new patterns will be loaded into the speech database. Depending on your processor speed, this could take anywhere from several seconds to several minutes.

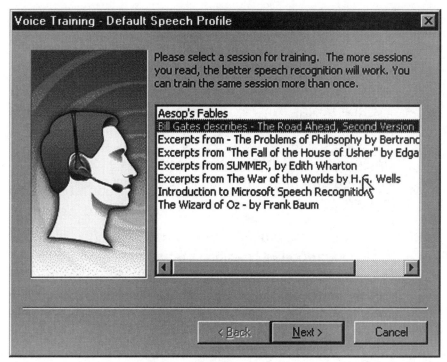

Figure 10.4

To reinitiate training later on, go to the speech toolbar, which is always active when speech recognition is active, and click on the **Tools** menu. Select **Training**. It is interesting to note that just because your copy of Office XP has learned how you speak, it has not learned the speech patterns of others. A new person sitting down at your computer will experience much lower accuracy because the system is trained for your use. There is a way, however, for multiple users — each with their own profile and training level — to use a single computer. We will show you how to set this up next.

Also, speech will work in other areas of Office XP, such as within Excel to enter data and move around cells. If speech will work within a component, you will be able

A new person sitting down at your computer will experience much lower accuracy because the system is trained for your use.

to activate it under the **Tools** menu. But, you will probably find the most use for speech-recognition technology within Word.

Many voices, many profiles

Just because Office XP is basically designed for a single user does not mean that you have to keep it that way. You can manage multiple users, but you may need to leave Office XP to do so.

When a new user wants to use your copy of Office XP to type reports or design spreadsheets and would like to use the speech-recognition engine, you will need to make a new profile. Otherwise, the system will be listening for your audio cues and will probably misinterpret the new users quite often.

To create a new profile, you need to go to the **Start** menu on the Windows operating system. Click on it and then select the **Settings** option. Then select **Control Panel**.

Once your Control Panel is up, find the icon that says **Speech** and double-click on it. This will bring up the speech window, as shown in *Figure 10.5*.

If you have not visited this screen before, there will be one item called **Default Speech Profile** in the window and it will have a checked box beside it. This is the profile that was created when the speech engine was initially installed.

To begin with a new user, click on the button labeled **New....** You will be prompted to type in the name of the

Just because Office XP is basically designed for a single user does not mean that you have to keep it that way.

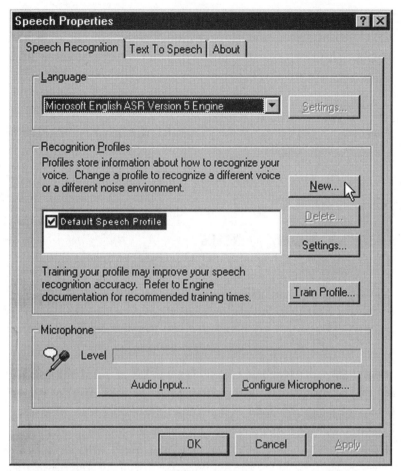

Figure 10.5

new user. Once this is finished, click **Next** > and the microphone adjustment wizard will launch. Once this is finished, the basic-training program that you went through when the engine was installed will run.

To delete a profile, select it and click on **Delete**.

Take note that whatever profile has the box to the left checked is the active one. The active profile needs to match the current user or else the recognition rates will

Take note that whatever profile has the box to the left checked is the active one.

be low. You can switch the current user from within Office XP if the speech toolbar is active. To do so, click on the **Tools** menu on the bar. Scroll down to **Current User** and select yourself.

Of course, if you are the only user on a system, you don't have to worry about managing multiple speech profiles.

When using English, the abbreviation EN is used.

Using Your Tools

Whenever you have speech activated, the speech toolbar, called **Language**, will be up. You can move this special toolbar around the screen, but it has an "always on top" property, so it won't lock into place beside another menu. You can minimize it if it is in the way. When minimized, it will drop into an icon on your main Windows taskbar. The engine will warn you of this however, as shown in *Figure 10.6*. The icon itself will be a little blue square with an abbreviation for whatever language you are currently using. When using English, the abbreviation EN is used. Double-clicking on the icon will bring the full menu back up.

The language bar is the main interface you will use to manipulate the speech engine. We already explained how to activate additional training sessions, but let's look at what else speech recognition can do.

It's best if you use a combination of keyboard and mouse clicks and voice commands. Things will get done a lot more quickly if you can juggle all these things, which is not as hard as it sounds.

Figure 10.6

Word has the most complicated speech toolbar. The one for Excel is fairly simple and consists of telling whether you are entering data into cells or directing the cursor to specific cells.

Word includes quite a few buttons. The first is called **Correction**. When you speak a sentence and the speech recognition engine produces the wrong words, you can use this button to fix it.

Highlight the mistake and then click on the **Correction** button. You will hear the phrase read back to you through your system's speakers or your headset speaker if you have one. Then the engine will give you a list of all things it thinks might be the right answer, basically the second- and third-best choices that, for one reason or another, the computer weeded out when picking the incorrect phrase. What you need should be in this list about 90 percent of the time. And remember, you can always type in what you want to say if the system ever stubbornly refuses to do what you desire.

You can always type in what you want to say if the system ever stubbornly refuses to do what you desire.

The second button is called **Microphone** and is basically an on and off switch for the speech-recognition engine. If the engine is active, the toolbar will be longer and will include the buttons **Dictation** and **Voice Command**. If you don't see these buttons, then the computer is not listening to you.

The long "balloon" on the toolbar will periodically update you on the status of the recognition rates.

Before you speak, click on **Microphone**. You will see the toolbar get longer and can begin talking to your system. As you work, if your phone rings or someone stops by or you simply need to leave, click on the **Microphone** button again to avoid having the engine try and transcribe a lot of conversations that are not supposed to be recorded.

The other two main buttons direct the engine to either listen for commands when the **Voice Command** button is pressed or to try and transcribe what you are saying, which is the default when the **Dictation** button is highlighted. So, when you want to save your work, click on **Voice Command** and say "Save." When you want to continue working, click on **Dictation** and start speaking. You only need to click the buttons once to activate their function. There is no need to hold them down like a walkie-talkie.

The long "balloon" on the toolbar will periodically update you on the status of the recognition rates. It will let you know when, for example, you are speaking too softly. If your recognition rates suddenly begin to drop off, glance at the speech balloon to get an idea of what might be going wrong.

All language bars throughout Office XP work the same way, with slightly different button commands depending on the program. But what is common to all of these applications is the **Tools** menu.

The **Tools** menu gives users the ability to tweak the engine if needed. The first command, **Learn from Document**, is quite handy. When you are finished working with a document, if you click on this option the engine will scan the text looking for any words that it does not recognize. If it finds any, you can add them to the custom dictionary and train the engine how you pronounce them. This is helpful if you work in a technical, scientific, or legal field, and use a lot of the same jargon that is not really included in the standard dictionary.

The **Options** menu takes you to the exact same menu we showed you how to get to from the Control Panel by clicking on the **Speech** icon.

Once there you can really tweak the engine, though it is not always advisable to do this. If you click the **Settings** button, you will come to another menu with two slider bars. The first adjusts the degree of pronunciation sensitivity the engine allows, though this won't affect your dictation, only the command and control functions such as when ordering the application to save a file.

The second slider bar lets you choose whether you want better recognition accuracy at the expense of speed or faster recognition at the expense of accuracy. By default, the slider bar is in the middle of the two extremes. In our tests, it's best if it stays there.

The final option in this window is that you can have the computer constantly adjust to the speaker, striving for greater accuracy. By default this is active and checked

The second slider bar lets you choose whether you want better recognition accuracy at the expense of speed or faster recognition at the expense of accuracy.

and we can't imagine why you would ever want to turn it off unless the PC's processor is so slow it can't handle learning-mode running in the background. But you should leave it on because the more you use the engine, the more the engine will know you and the higher the accuracy rates will climb.

The **Show Balloon** option turns on and off the little text balloon status indicator. If you really don't like it, you can turn it off here.

The **Training** button activates more training sessions. Although it is best to do different sessions, your accuracy will also improve if you do the same ones over again.

Add/Delete Words(s) lets you examine all those nontraditional words that the engine has added to the custom dictionary and listen to what the computer thinks they sound like. You can delete words, add new ones, or fix the ones that the computer thinks sound totally different from how they actually sound. Here you can increase your accuracy rate, one word at a time.

We already told you how to use the **Current User** option. That pretty much is all there is to know about speech recognition within Office XP.

Don't talk back?

An interesting feature is that in certain circumstances, your computer can actually talk back to you. But don't worry, it will only say what you tell it to.

When you are using Excel, go to the **Tools** menu and select **Speech** and then **Show Text to Speech**

You can delete words, add new ones, or fix the ones that the computer thinks sound totally different from how they actually sound.

Toolbar. This will bring up a very tiny toolbar that can be used to activate the text-to-speech function. If the feature was not installed, you may have to insert your main Office XP disks to complete this process.

Once you have the toolbar up, you will notice that you only have five buttons. The first is a little green triangle, like the play button on a VCR. Pressing this will, not surprisingly, activate the speech engine. You will need to have something typed into the Excel spreadsheet or it won't say anything.

The second button is a red square and this causes the engine to stop reading text in case you have started the computer reading a very long document and need to stop.

Once you have the toolbar up, you will notice that you only have five buttons.

The third and fourth buttons are the Rows and Columns button. This tells the computer whether you would like it to read cells horizontally — by rows — across the page or vertically — by columns. The diagram of the arrows shows which way is which, in case you get confused.

The final button is the "Speak on Enter" button. This button makes the speech engine say whatever is in a cell as soon as you hit Enter. So each time you enter data, you will hear whatever you have just typed. This can be a huge help if you are entering a lot of numbers and would like an audio confirmation of each entry as you make it.

One thing to keep in mind when using the speech tool is that you need to have a cell highlighted within the block of text and numbers somewhere to activate the system. If you have a cell highlighted in the middle of a group of data when you push play, Excel will highlight all the data and then proceed to read either by rows or columns.

If, however, you have the highlighted cell off by itself somewhere, not within the data or one cell beside or above or below it, the engine will not highlight anything and will not begin reading.

If you would like to have the program read just a small sampling of data within a larger field, highlight just what you want it to read and press play. It will only automatically highlight all the data if you have just one cell highlighted when you press play.

There are a few limited controls to make the voice read faster or slower, or even change gender. The best way to access these controls is by going to your control panel outside of Office and selecting Speech. The center tab in the control menu that will appear should say Text to Speech, as shown in *Figure 10.7*.

You only have a few options. The first lets you select whether you would like to have a male voice named Michael or a female voice named Michelle read your text. Pressing the **Preview Voice** button will let you hear whatever voice option is selected. Be sure to hit the **Apply** button to make your changes take effect.

You can also change the voice speed by moving the slider bar in the menu left for a slower voice or right for a faster one. Hitting the **Preview Voice** will give you a sample of the new voice speed, so you can configure it as you like.

There are a few limited controls to make the voice read faster or slower, or even change gender.

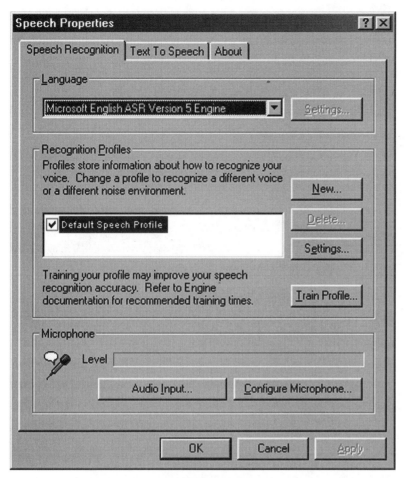

Figure 10.7

The only other option is **Audio Output,** which most users won't have cause to use. If you have multiple soundcards or audio devices, you can select which one the system should use to output the voice. However, if the system is working fine, it would be advisable not to mess with these controls as 98 percent of desktop computers only have a single sound device anyway.

If you have multiple soundcards or audio devices, you can select which one the system should use to output the voice.

If you work at it and take the time to train your system, you will be rewarded with nearly hands-free use of the Office XP suite in addition to getting your work done a lot faster.

Keep at it

Finally, don't get frustrated. Remember that your computer can't see you and that Office XP does not know you and your speaking patterns right off the bat. Truly high recognition rates require a lot of training time with your system, though much less than when processor speeds were slower. On a Pentium II, training times for one session with some engines took more than three hours, plus another 40 minutes to crunch the data once it was complete. The Pentium III chip reduced the average training time down to just 15 minutes per session followed by a minute of data crunching. The Pentium 4 brings that time down again.

If you work at it and take the time to train your system, you will be rewarded with nearly hands-free use of the Office XP suite in addition to getting your work done a lot faster. And that is something that would make even a science-fiction hero jealous. So don't be afraid to "Engage!" your speech engine and take time to let Office XP get to know you.

11 The Office Web Site

Even with all the user-friendly enhancements with Office XP, you are still going to run into a time when you need something you don't have or just want to make use of online resources. Most of the additions that can be downloaded to improve Office are free, so there is no reason not to get them if you have the time and space on your computer. Some of the security updates could even end up saving your system.

Most of the additions that can be downloaded to improve Office are free.

Sites at Microsoft.com

Thankfully, Microsoft does a good job of supporting Office XP. You can start at http://www.microsoft.com/office. This site can serve as the launching point if you need specific information about the suite. You can get there by typing the address into your browser.

From this basic information area you can see if there is anything new about Office and purchase nonstandard Office XP family applications if you wish.

If you see any security updates, it is highly recommended that you download them right away.

Eventually, if you want to download a support file, you may be taken to Microsoft's Office Update site at http://officeupdate.microsoft.com. You can also automatically go to the page by selecting the **Office on the Web** option under the **Help** menu. Once you arrive at this page, a quick scan should tell you of all the different updates that have come out since you installed Office. Clicking on the link should start the download process and, in most cases, the update will install itself on your system, as shown in *Figure 11.1*.

Some of the most important updates are in the field of security. If you see any security updates, it is highly recommended that you download them right away.

Office is a very popular suite for both business and home users and the malicious folks who write viruses know

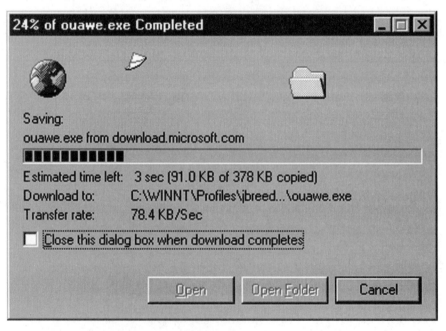

Figure 11.1

this. Because of this, a lot of virus code is written specifically to take advantage of the Office suite. Attachments that come in through e-mail sometimes try, for example, to get into your Outlook address book and then e-mail themselves out to all your contacts; this is known as a *self-propagating virus* or a *worm*. You might have heard of some of the more prolific viruses or worms by the names *Melissa*, *Anna Kournikova*, and the *ILOVEYOU/LoveBug/LoveLetter*.

Other malicious code takes advantage of the macro options in many Office applications. Macros allow you to automate by recording tasks and playing them back later. Microsoft's underlying code is somewhat powerful. Some malicious folks tap into this code and create macro viruses specific to Word, Excel, or even PowerPoint. Someone might send you a document that's infected and could infect your computer.

Macros allow you to automate by recording tasks and playing them back later.

Office XP warns you if any hidden code is trying to send out mail on your behalf or if some macros are trying to execute without your knowledge. Under the **Tools** menu, select **Macro** ▸ and then **Security...**, as seen in *Figure 11.2*. We recommend at all times to set your security to **High**.

However don't expect virus writers to rest until they have found ways around this. So, keep checking for updates, just in case. Microsoft makes every effort to provide security, but dozens of new viruses come out each week and it is sometimes necessary to get updates.

While Office XP tries to protect itself, we find a more comprehensive approach to security is necessary. We suggest purchasing anti-virus software like Symantec Corp.'s Norton AntiVirus or Network Associates Inc.'s McAfee VirusScan.

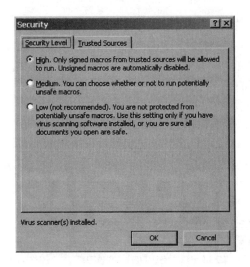

Figure 11.2

The Office XP Web site provides more than product updates.

Another service that you can sign up for off the Web page is the Office Auto Update Notification Service. You basically give Microsoft your contact info. Microsoft sends an e-mail message to you whenever an update to Office is available. We have used this service in previous versions of Office and found it helpful, especially if you have lots of applications you are managing and don't have time to search the Web pages to find each one.

The Office XP Web site provides more than product updates. A lot of helpful information about Office is available. Several tips and tricks are posted at the site along with files to perform new tasks you never thought about.

Design Gallery Live!

Not all the interesting things that come with Office are found by poking around the Web. Some things are embedded right in the applications.

The best example of this is Design Gallery Live, formerly called Clip Gallery Live. When you search for clip art in any Office XP application, you will find that there is a fairly extensive library included on Office CD-ROMs. But no application can have every piece of clip art you might ever want. Microsoft keeps a large variety of clip art online. The beauty of this is that since you are accessing several servers, the space for clip art is pretty unlimited. Also, because of the interface, it's practically seamless. And once you download new art, it is then stored on your hard drive for instant use from that point on.

To get to the Online Gallery, go to the **Insert** menu and select **Picture** ▸ and then choose **Clip Art....** The screen in *Figure 11.3* should look pretty familiar if you read the chapter on Word.

When you search for clip art in any Office XP application, you will find that there is a fairly extensive library included on Office CD-ROMs.

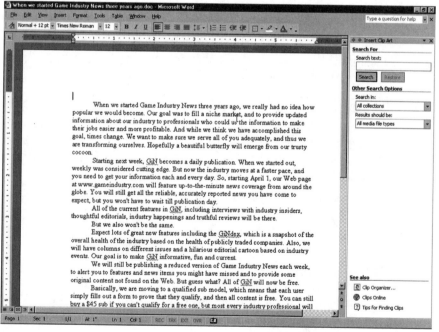

Figure 11.3

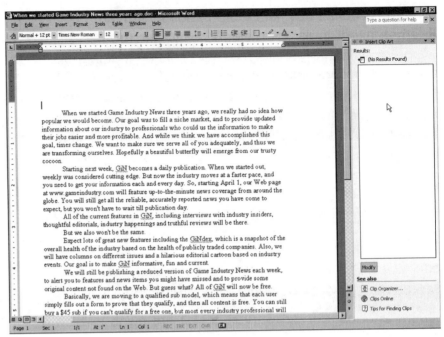

Figure 11.4

You could go surfing out on the Web, but first, let's try the live gallery.

Let's search for something that the included database does not have. Try **penguin**. What, no penguins?! What do you do if you are shown the message in *Figure 11.4*?

Don't panic. Just because it did not come on the Office CD-ROMs does not mean that there are no penguins anywhere to use. You could go surfing out on the Web to try and find them, but first, let's try the live gallery. At the bottom of *Figure 11.4*, you will notice it says **See also** and then there are three options.

The first is **Clip Organizer...**. This is where your clips are stored locally, either on a CD-ROM or hard drive. Those clips have already been searched.

Another selection is the **Tips for Finding Clips**. This basically launches the **Help**, though it is specifically tailored in this case to show you search techniques.

The option that you want to select is the **Clips Online**. When you select this, you will be taken to the Web site where you can look for the appropriate clip. If you are offline, your Internet connection window will appear and you will have to type in your password and account information. If you have a direct connection to the Internet, the passage to the online clip gallery will almost be seamless.

Once you get there you will be shown a search engine as in *Figure 11.5*. Simply type in what you want to find, in this case **penguins**, and hit enter or the **Search** button. You can specify that you are looking for clip art or photos, and if you care what type of media is found. You can even search for sounds or movie clips, depending on what type of document you are creating.

You can specify that you are looking for clip art or photos, and if you care what type of media is found.

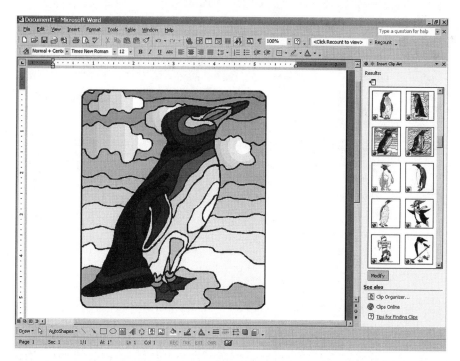

Figure 11.5

*Each
downloaded
piece of art is
stored in your
personal
gallery.*

Once you find what you need — look at all the penguins — you can click on them to download the art into your hard drive. Even with a 56-kilobit-per-second dial-up connection, it should go pretty quickly because the files are so small. Each downloaded piece of art is stored in your personal gallery, which you can access through the **Insert Clip Art** Task Pane.

If you've got a persistent connection to the Internet — like a cable modem or a digital subscriber line (DSL) — and Office notices this, you may not have to visit the Web site.

When you type in **penguins** in the **Insert Clip Art** Task Pane, Office could go out and search the Design Gallery Live Web site automatically.

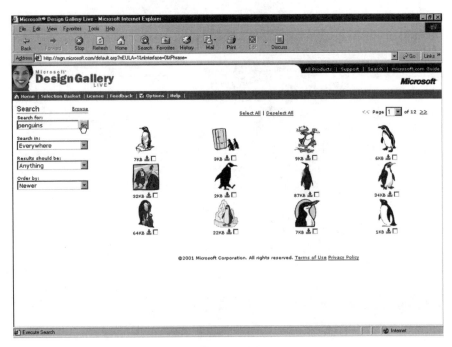

Figure 11.6

Notice within the **Insert Clip Art** Task Pane that some little globes appear in the bottom left corner of all penguin images as in *Figure 11.6*. That means Office downloaded the penguins from the Web.

Template mania

The new Office provides a lot of templates. A template exists for just about anything of which you can think.

A template exists for just about anything of which you can think.

However, remember that everyone is going to have access to those same templates, so if you are not careful, your resume, fax, or letter might end up looking just like someone else's. Plus, there is always an exception to the rule. Perhaps you need a specific type of document and there is not an included template with the suite.

There is hope.

When you are in an application within Office XP such as Word or Excel and want to load a template, you first go to the **File** menu and click on **New**. This will open a Task Pane.

The final section in the Task Pane is **New from template**. If you click on **General Templates**..., you will be shown the standard templates that come on Office XP CD-ROMs, much like when you select **New Office Document** from the Windows Start menu. The second option **Templates on my Web Sites...** will take you to storage areas with templates that you have set up or provided by your company if you work in an office. You can add Web folders to this list by clicking on the **Add Web Folder...** or the **Add Network Place...** link at the very bottom of the Task Pane.

Most users, however, will not have access to or know of a Web page devoted to templates. Luckily, Microsoft has a bunch of ready-to-use templates at the Microsoft site.

You can find and download templates off the Office support Web sites without actually going into an Office application.

Simply click on the **Templates on Microsoft.com** option to get access to the templates. Here you will find a lot of new variations on old themes, like different-looking resumes, and also a few special templates that are not part of the original program. Clicking on the template will load it onto your open application. Again, if you are connected to the Internet, this process will be seamless. If you are not, then you will have to enter your dial-up or connect information and passwords.

Browse until you find the template you want — you will have to accept a license agreement first — and then when you click on **Edit in Word**, it will be downloaded and put into Word for you. Of course, other applications offer their own templates, including Excel and PowerPoint.

It is interesting to note that you can also find and download templates off the aforementioned Office support Web sites without actually going into an Office application, though it is generally easier to use these helpful features inside Office XP itself.

Doctoring Office XP

Even the healthiest person sometimes gets sick. Likewise, even the healthiest program can become corrupted under the right, or perhaps the wrong, circumstances. Viruses can attack, of course, but things can also go awry if certain events occur, like improperly shutting down the computer while Office XP is open — such as in the case of a sudden power outage.

The support page we showed you at the beginning of this chapter will go a long way toward keeping Office XP running. You can also use the Windows Update option under the Start menu to find updates that are ready for Office. In our tests, the operating system did a good job of detecting Office and steering us toward updates when using this function.

If, however, you can't seem to find an update to fix an Office XP application, you also have options. One is less drastic than the other and should be attempted first.

The first thing that can be done is a good option if you can launch an Office XP application, but it appears to be not acting correctly.

Go to the **Help** menu and select **Detect and Repair...**. A **Detect and Repair** window will pop up. The first option that is checked lets the application restore any broken shortcuts. It's good to leave this checked. The second option lets Office disregard all your customized settings and replace them with the defaults. At first, leave this option alone unless you suspect that somehow your customized settings led to the broken Office. This is not very likely. Whatever you choose to do, click **Start** to continue.

Office XP will then begin a self-diagnosis of all the different files, checking to see if any of them are damaged or corrupt. You will likely be asked to insert your installation CD-ROMs during this process. You may also

You can use the Windows Update option under the Start menu to find updates that are ready for Office.

be asked to close any open applications running on your computer, as these might interfere with the diagnosis and fixing process.

Even on a very fast computer, expect to wait a little while for Office XP to check itself out.

Even on a very fast computer, expect to wait a little while for Office XP to check itself out. It has to examine each and every file for errors, and there are thousands. When it is finished, it will either tell you that nothing is wrong or attempt to fix problems.

If this method does not work or you can't get into any of your Office XP applications to launch the repair routine, then your other option is to insert your installation CD-ROM again. When it Autoruns, select that you want to perform a refresh install. Specific details about how to accomplish this if you get stuck are in the chapter on Installation. This will replace all the files and should fix your problem. If it doesn't, then your problem is likely with the computer itself or the operating system and not Office XP.

When you either fix or reinstall Office XP, no data files are deleted or altered in any way. Completed documents will not be touched by either the repair or fixing process. The only way the data files themselves will be harmed is if they have been targeted by a virus. While the security updates should help prevent this, it is always a good idea to have a virus scanner on your computer and to update the virus definitions for the scanner monthly.

Strength in numbers

The Web support for Office XP should keep most users on track with updates and product fixes as they become available. Plus, the templates and clip art add a huge library of information for the taking. Don't be afraid to use these features. You bought the right to them when you purchased Office XP. And some of the templates and clip art are pretty fun. Explore them and increase your productivity at the same time.

The templates and clip art add a huge library of information for the taking.

12 Working with the Web

While Chapter 11 explained how Microsoft's Web site helps Office XP, this chapter takes Office to the next level.

Office XP embraces the Web like no other Office ever before. More than turning Web addresses into links, Office XP can use the Web dynamically.

How the Web Integrated with Office

Microsoft made HyperText Markup Language — better known as HTML — an optional format for Office applications back with version 97. Office 2000 worked somewhat better. HTML as a file format is available for almost all circumstances for all Office XP applications — from an e-mail to a database.

More than all that, Office XP takes full advantage of the Web. Let us show you.

HTML as a file format is available for almost all circumstances for all Office XP applications.

Open up Excel 2002 and a blank worksheet. From the **Data** menu, select **Export External Data ▼**. Then choose **New Web Query…**.

Remember, in Chapter 8 on FrontPage 2002, tables on Web pages are somewhat similar to the cells in Excel. Excel's Web query will look for tables on a Web site. You need to know a Web site that contains a table, perhaps with financial data.

How about http://www.exchangerate.com (*Figure 12.1*)? It keeps information about international monetary exchange rates. But it can be any site. If you know of one that maintains information on your stocks, bonds, or mutual funds and presents it in a table, you can query it, too.

Tables on Web pages are somewhat similar to the cells in Excel.

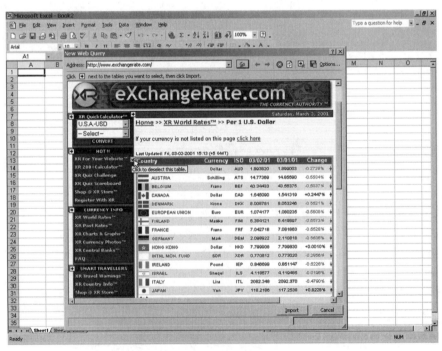

Figure 12.1

From the **New Web Query** dialogue, enter the address you want to visit, as seen in *Figure 12.1*. Select the appropriate table, turning the arrow icon into a check mark. The **Options...** button in the top right corner allows some control over how the information is imported and whether it keeps the HTML formatting or just the data. Note that it will not import graphics, like the little icons of country flags.

Then click the **Import** button. Excel will ask whether to create a new worksheet or insert the table into the current worksheet and at what cell. Bring it in to $A$1 and click **OK**.

Excel will ask whether to create a new worksheet or insert the table into the current worksheet and at what cell.

The resulting query brings in all the data from that site's table. The **External Data** toolbar opens up, too, as seen in *Figure 12.2*.

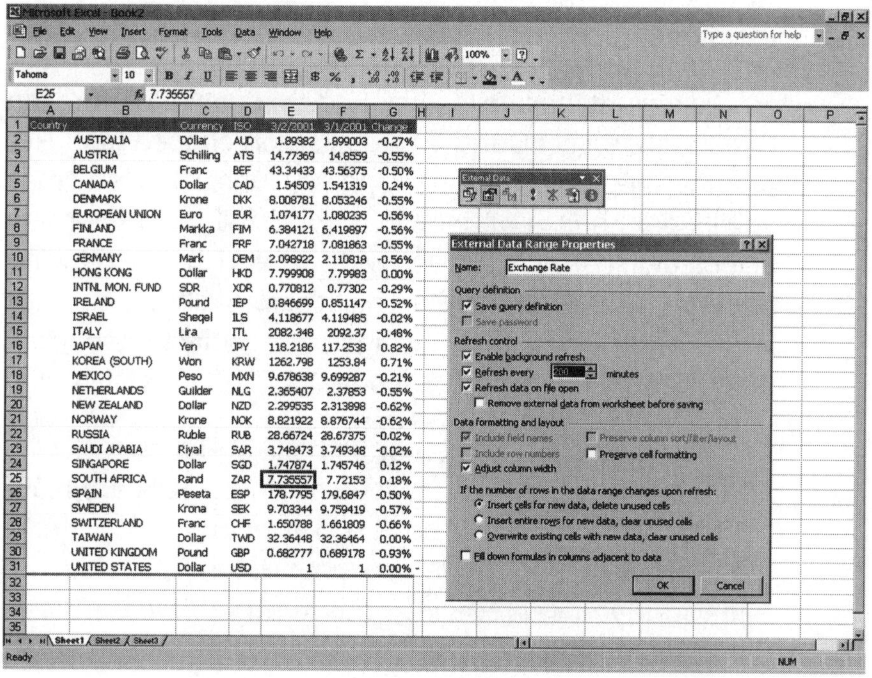

Country	Currency	ISO	3/2/2001	3/1/2001	Change
AUSTRALIA	Dollar	AUD	1.89382	1.899003	-0.27%
AUSTRIA	Schilling	ATS	14.77369	14.8559	-0.55%
BELGIUM	Franc	BEF	43.34433	43.56375	-0.50%
CANADA	Dollar	CAD	1.54509	1.541319	0.24%
DENMARK	Krone	DKK	8.008781	8.053246	-0.55%
EUROPEAN UNION	Euro	EUR	1.074177	1.080235	-0.56%
FINLAND	Markka	FIM	6.384121	6.419897	-0.56%
FRANCE	Franc	FRF	7.042718	7.081863	-0.55%
GERMANY	Mark	DEM	2.098922	2.110818	-0.56%
HONG KONG	Dollar	HKD	7.799908	7.79983	0.00%
INTNL MON. FUND	SDR	XDR	0.770812	0.77302	-0.29%
IRELAND	Pound	IEP	0.846699	0.851147	-0.52%
ISRAEL	Sheqel	ILS	4.118677	4.119485	-0.02%
ITALY	Lira	ITL	2082.348	2092.37	-0.48%
JAPAN	Yen	JPY	118.2186	117.2538	0.82%
KOREA (SOUTH)	Won	KRW	1262.798	1253.84	0.71%
MEXICO	Peso	MXN	9.678638	9.699287	-0.21%
NETHERLANDS	Guilder	NLG	2.365407	2.37853	-0.55%
NEW ZEALAND	Dollar	NZD	2.299535	2.313898	-0.62%
NORWAY	Krone	NOK	8.821922	8.876744	-0.62%
RUSSIA	Ruble	RUB	28.66724	28.67375	-0.02%
SAUDI ARABIA	Riyal	SAR	3.748473	3.749348	-0.02%
SINGAPORE	Dollar	SGD	1.747874	1.745746	0.12%
SOUTH AFRICA	Rand	ZAR	7.735557	7.72153	0.18%
SPAIN	Peseta	ESP	178.7795	179.6847	-0.50%
SWEDEN	Krona	SEK	9.703344	9.759419	-0.57%
SWITZERLAND	Franc	CHF	1.650788	1.661809	-0.66%
TAIWAN	Dollar	TWD	32.36448	32.36464	0.00%
UNITED KINGDOM	Pound	GBP	0.682777	0.689178	-0.93%
UNITED STATES	Dollar	USD	1	1	0.00%

Figure 12.2

Office XP's integration with the Internet continues in other areas.

The first button on the toolbar, named **Edit Query**, takes you back to the **New Web Query** dialogue. The second button is rather powerful. The **Data Range Properties** dialogue, as seen in *Figure 12.2*, can control how often the information in the spreadsheet is automatically refreshed. Set it to 0 (zero) minutes and it refreshes constantly. Or have it refresh just when the file is opened. You can also determine how you want new rows or columns handled.

The buttons **Refresh Data** and **Refresh All** can check the Web for updates at your request just by hitting one of those buttons.

Office XP's integration with the Internet continues in other areas as well. Say you want to send a portion of an Excel spreadsheet. You don't want to send the whole workbook — just a selection of cells appropriate for the recipient.

Highlight the section you want to send and, from the **File** menu, select **Send To ▶** and then choose **Mail Recipient**. The Office Assistant appears (one of the few times you'll see the little paper clip or whatever character you selected) to ask whether you want to send the entire workbook as a file attachment or just the worksheet currently in use as a part of the message body. Select the current worksheet in the message body.

The Excel sheet then appears like *Figure 12.3* with a section like an e-mail. As long as some cells are highlighted, the option will be **Send this Selection**. That means only the highlighted portion will be included in the e-mail, as you can see was delivered in *Figure 12.4*. If you'd prefer to send the whole worksheet, just click a single cell and the choice turns to **Send this Sheet**.

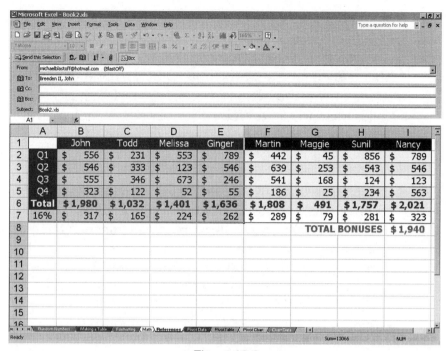

Figure 12.3

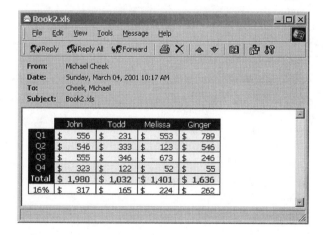

Figure 12.4

While all the applications in Office XP can send copies of themselves from the **File** menu, only Excel 2002 allows this selection option.

What is XML?

Any discussion of the Web today has to include XML. XML stands for eXtensible Markup Language. XML puts structure onto the Web. We've been over HTML, the code that is used to make Web pages appear as they do. XML isn't as much about appearance as much as it is structuring the data.

XML stands for eXtensible Markup Language.

Figure 12.5 shows the raw XML. The XML file contains structured data for both content and tags. The tags contain information to note what role the associated content plays. Content can be words or pictures or any other data available to the Web. The tags can, for example, note the difference between a section heading versus a footnote or it can note the relationship between an im-

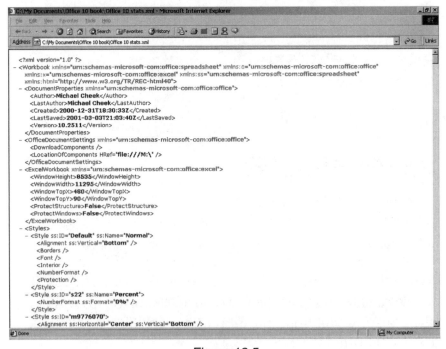

Figure 12.5

age and its caption. Almost all documents have some structure. While HTML can be freeform, XML is not. That's not to say XML isn't flexible; it can conform to most documents.

XML provides a much stronger ability for exchanging information and works from Web-based connections to large databases.

That said, it's very early in XML development. The current version of XML is 1.0, which shows just how early.

Microsoft Office XP includes XML support.

Microsoft Office XP includes XML support. In fact, some have suggested that the reason Microsoft chose XP as the moniker wasn't "eXPerience," but as a hint to its initial support for XML.

Two Office XP products support XML — Excel 2002 and Access 2002. Since both use structured data, it's relatively easy to provide some support. Moreover, the Web query discussed earlier in this chapter can tap into XML data incorporated in tables.

What is SharePoint?

You might have heard about Intranets. The Intranet provides resources within an organization to exchange information. Intranets work a lot like the Internet, only the audience is much smaller and more focused. For larger companies, Intranets can be quite handy. But, sometimes smaller teams work together to accomplish similar goals. Such small teams may be clustered in a cube farm or across the world from one another.

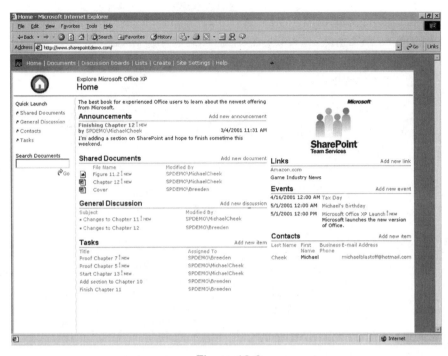

Figure 12.6

SharePoint provides a platform for sharing information, documents, calendars, and more.

Microsoft has built a new Intranet shell called SharePoint Team Services. Ideal for teams working on projects together, SharePoint provides a platform for sharing information, documents, calendars, and more. *Figure 12.6* shows how we worked together on the book in a SharePoint Web site.

SharePoint includes the following components:

♦ **Announcements:** List team events, accomplishments or just a little news from someone's successful promotion, to critical data you want team members to see first thing.

♦ **Documents:** Files can be saved directly to SharePoint or uploaded while visiting the Web site. Users can even sign up to be notified by e-mail when a document is changed. This can be an ideal place to deposit often-used documents like expense report forms and time sheets.

♦ **Discussions:** Start a public discussion on an issue and get feedback directly from the team. This open forum provides a free flow of ideas and keeps clutter out of the inbox. E-mail doesn't have to be cc:ed to everyone on the team.

♦ **Tasks:** Similar to the Tasks in Outlook, this forum is more ideal when the whole team is not on the same network. Individual tasks can be "exported" directly into Outlook or the whole group can be imported into Excel as a Web query and updated regularly. Users can also subscribe to tasks for e-mail updates.

♦ **Events:** Just like Tasks, Events can be imported individually into Outlook or as a group into Excel. Users can subscribe for e-mail updates. SharePoint also includes a group calendar that shows the month's events on a single screen.

♦ **Contacts:** Just like Events and Tasks, Contacts allows Outlook users to send any contact to SharePoint. Of course, contacts can be imported from SharePoint into Outlook as well.

SharePoint also automatically indexes content so text searches can be performed on all the contents in the site.

User management and permission levels can be performed from the Web interface, as seen in *Figure 12.7*. The components available and basic layout can also be managed from a Web interface. SharePoint is pretty easy to understand, navigate, and manage.

Just like Tasks, Events can be imported individually into Outlook or as a group into Excel.

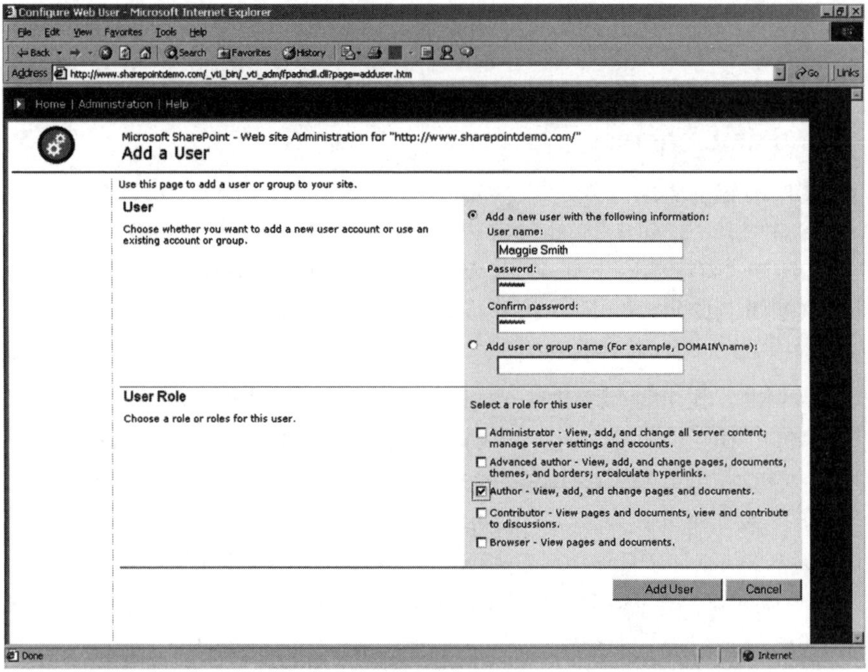

Figure 12.7

SharePoint also can incorporate a "sub-Web."

SharePoint also can incorporate a "sub-Web," meaning a portion of the site can be totally customized and authored by a member of the team. Moreover, using FrontPage, the appearance of SharePoint can be altered to another theme.

Managing the SharePoint Web site — or any Web site, for that matter — can be relatively easy and integrated into using Office XP and the Windows operating systems.

Integrating the Web

In the operating system, go to the **Network** or **Network Places** icon located on your desktop. For Windows Me, it's called **My Network Places**. Launch it and select **Add Network Place**.

A wizard pops up like in *Figure 12.8*. Just type in the Web address and don't forget the **http://** at the beginning. You'll be asked for your username and password, but easily enough, the site appears as part of your **Network Places** — like a hard drive mapped to somewhere else on the network. You can save files to the site as well. If you have rights, you can access it.

The site appears as part of your **Network Places**.

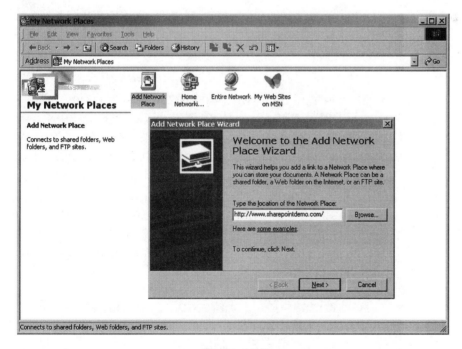

Figure 12.8

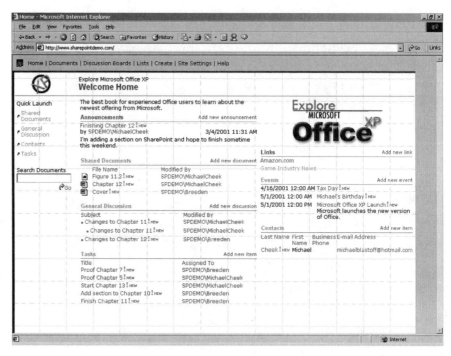

Figure 12.9

In FrontPage, you can open SharePoint just like you open a regular Web site.

In FrontPage, you can open SharePoint just like you open a regular Web site. From the **File** menu, select **Open Web...** then choose it from the **Network Places**. In FrontPage, you can edit graphics or even change the theme, as we did. Check out *Figure 12.9* for our changes, compared to *Figure 12.6*. This allows teams to build sites appropriate to them. Content can change dynamically as appropriate.

Security Issues

As information is shared on the Web and across the Internet, security matters more and more. Microsoft has thought of this and integrated more security features in Office XP.

In Chapter 11, we discussed macro security as well as viruses. Again, we recommend being very careful about using macros since thousands of Microsoft Office macro viruses are out there and more are created every day. Maintaining your software updates and vigilant anti-virus efforts assures you of operating in a safe environment.

Thousands of Microsoft Office macro viruses are out there and more are created every day.

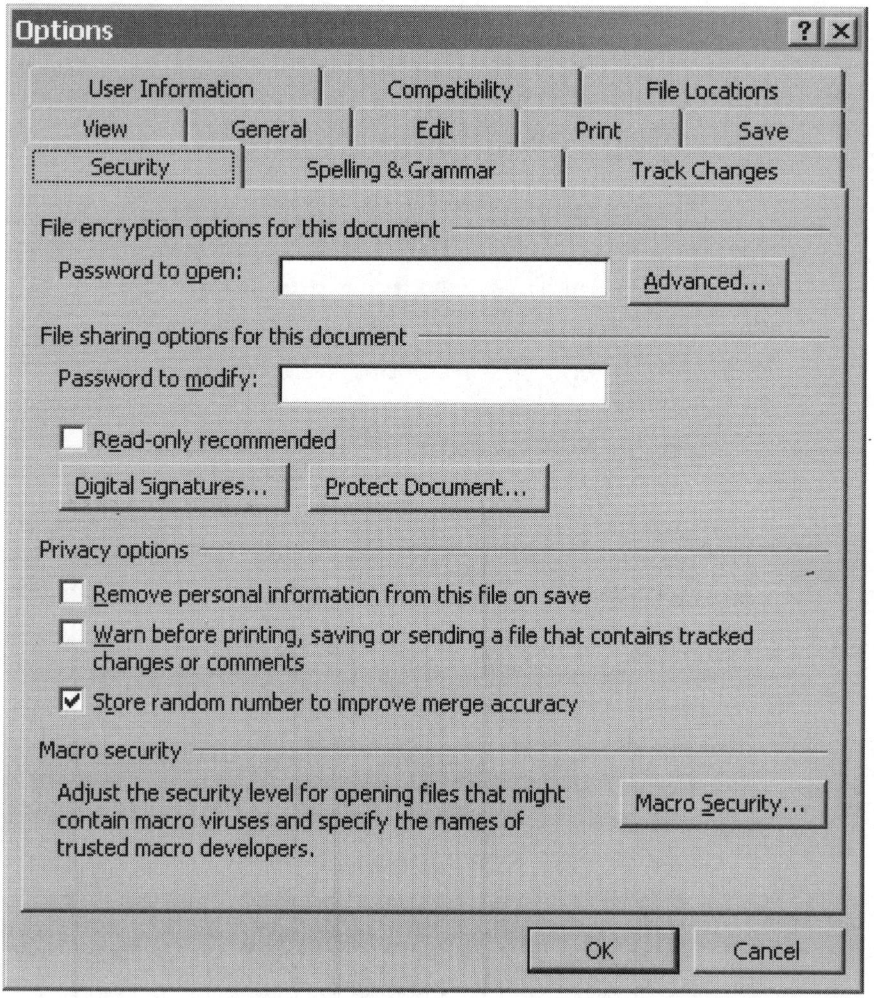

Figure 12.10

Office XP offers a few more security tools for additional defense. Password protection, encryption, digital signatures, and privacy options are also available for Office XP documents. You can access these features from the **Tools** menu. Select **Options**… and then choose the **Security** tab (*Figure 12.10*).

Keep in mind that no security efforts are complete without a comprehensive approach, including physically preventing unauthorized personnel from accessing the computer itself. These new Office XP components are only a link in a chain of security measures.

Password protection has been available for Office documents for a while, but now those passwords incorporate encryption of the entire document. Next to the **Password to open:** field, click the **Advanced…** button to choose an encryption option, like those in *Figure 12.11*. The maximum complexity is 128-bit, which is generally strong enough as a deterrent.

> *Password protection, encryption, digital signatures, and privacy options are also available for Office XP documents.*

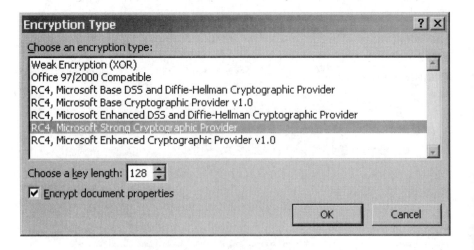

Figure 12.11

You can also restrict people's rights for modifying the document by creating a password so only those people with that secret code can make modifications. Otherwise, unauthorized people can view the file. Just type it in **Password to modify:**. The **Protect Document...** button is similar. As the author of a document, you can permit reviewers to insert just comments — meaning the content cannot be altered — or actually make changes. Choosing **Tracked changes** gives full control. Choosing **Comments** only allows the notes to be added and no changes can be made to a document.

Documents can be signed with a digital signature. This signature is a verification tool to determine if someone is legitimately the author of a document. For example, if John and Beth are working together on a top-secret project, they might exchange digital signatures and "trust" each other. Whenever John receives a document from Beth, he can check the signature to assure that Beth is actually the author. Macros can also be signed in this way. Click on the **Digital Signatures...** button to attach a certificate to a document and "sign" it. Such certificates are generated by companies like VeriSign Inc. or company-wide by a security manager, so a digital signature may not be an option.

The **Privacy Options** removes some personal information and notifies if a file is sent, saved, or printed. You can also access the **Macro Security...** options here, which were discussed in Chapter 11.

While security is provided, keeping your computer and its files out of the hands of the malicious among us requires vigilance. Strong passwords that use numbers, upper- and lowercase letters, and odd characters always

Documents can be signed with a digital signature.

Avoid using words in any dictionary or people's names.

help. Your PC should have such password protection. Avoid using words in any dictionary or people's names. Several applets are available that attempt what's called a "brute-force attack." These applets go through dictionaries and repeatedly attempt passwords.

Security is in your hands. The weakest link should never be you.

13 All Together Now

Each application in Office XP received quite a tune-up from Microsoft — from major changes like Smart Tags and Task Panes to more subtle changes like keeping that little paper clip from appearing automatically when you start using any Office product.

The way Microsoft Office XP applications work together has improved dramatically, too. The applications blend together, more intertwined than ever before, to thunder like an orchestra or harmonize gently like a folk duet.

The way Microsoft Office XP applications work together has improved dramatically.

Harmonizing with Smart Tags and Artificial Intelligence

One new feature of Office XP is Smart Tags. Throughout this book, we've highlighted how Smart Tags work. But let's see how Smart Tags might work with a little help from some artificial intelligence.

Open up Word 2002 and type a name, perhaps of a friend or someone famous. Type today's date. How about a local street address and maybe a telephone number?

You might notice some or all of these elements get a light purple underline. If not, you need to turn on some Smart Tag options.

Go to the **Tools** menu and select **AutoCorrect Options...**. Choose the **Smart Tags** tab, which should look something like *Figure 13.1*. Make sure **Label text with Smart Tags** is checked and then choose which types of text you want Word to highlight. You can click **Check** (or **Recheck**) **document** when you're done. You'll get a warning that you might lose some Smart Tags, but go ahead and click **Yes**.

> *You might notice some or all of these elements get a light purple underline.*

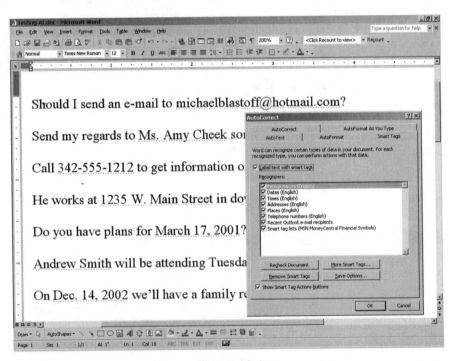

Figure 13.1

Now the purple, dashed underlines should appear under people, places, and dates — practically anything you've selected.

Microsoft will be offering new Smart Tags as will some third parties. To check out additional offerings, click the **More Smart Tags...** button and you'll visit the Microsoft Web site for details.

Each type of Smart Tag is different. As the cursor hovers over each item, its Smart Tag appears. Click on the ① Smart Tag icon and the drop-down menu provides appropriate options, as seen in *Figure 13.2*. For example, in regard to a person's name, the Smart Tag menu offers to send an e-mail message using Outlook 2002, schedule a meeting in Outlook, open the contact

As the cursor hovers over each item, its Smart Tag appears.

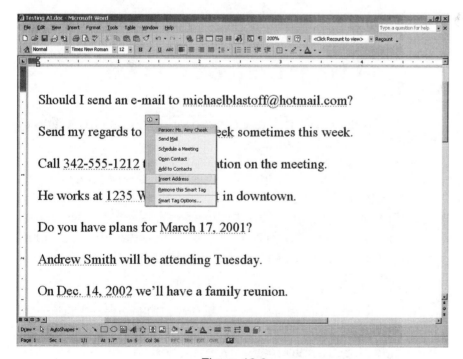

Figure 13.2

information from Outlook, add a contact entry in Outlook, or insert the address from the Outlook contact into the Word document. For a date, the Smart Tag options include scheduling a meeting in Outlook or opening the Outlook Calendar. Addresses also link into Outlook, but can also go to the Web to show a map of the location or get driving directions.

Excel offers limited Smart Tags using artificial intelligence.

If your PC is a little slow or you don't have as much memory, you might want to turn this off. Still, it can come in handy.

Excel offers limited Smart Tags using artificial intelligence, mostly restricted to e-mail addresses from Outlook. The other Office XP applications do not offer such Smart Tags.

And What Outlook Can Do...

While Outlook 2002 lacks the Smart Tag references available to Word, it does incorporate some strong ties to bring Office closer together.

Just as Word ties into Outlook, Outlook ties somewhat back into Word. Open Outlook 2002 and select the **Contacts** list. From the **Actions** menu, choose **New Letter to Contact**. You can choose this from the main Outlook application or from an individual contact, as in *Figure 13.3*.

Once this is selected, Word 2002 launches directly into the **Letter Wizard**, as seen in *Figure 13.4*. Word inputs all of the information from Outlook automatically. Just

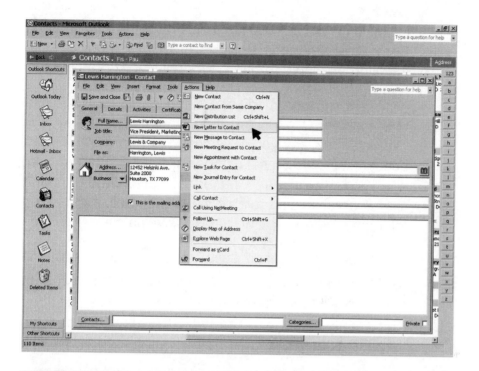

Figure 13.3

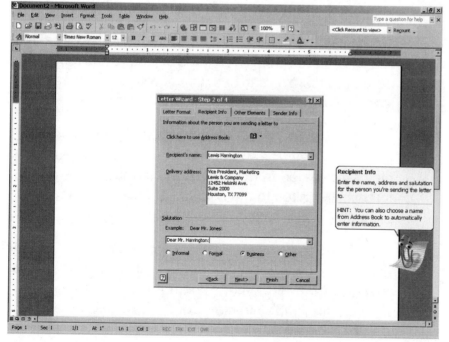

Figure 13.4

by making the appropriate adjustments — indicating a **Formal**, **Business**, or **Informal** letter, for example — and pressing the **Next >** button, Word builds the letter.

That **Actions** menu offers a lot of options to link information together in a more comprehensive way. For example, let's say you're working hard on several sales accounts. Sometimes it can be difficult to recall the work you've done on each account. Sure, with the long filenames or saving related items together in the same folder, you can keep abreast of individual files. A search of your Inbox will net all e-mail messages you sent to a client. You might also want to search the Sent Items for all of the e-mail messages you sent. You might also want to search through all your appointments in the Calendar. And it might be helpful to see who that client works with in your Contacts database.

Okay, as you can see, this can get pretty overwhelming.

Using the **Actions** menu along with the **Journal** feature in Outlook 2002, everything can be linked together in a single place. You can better track your e-mail messages, letters, PowerPoint presentations, and more.

In *Figure 13.5*, some of the items that can be linked together appear. If e-mail messages, appointments, tasks, etc., are created from the **Actions** menu, the item is automatically linked to the contact. The **Activities** tab on a Contact's card contains all of the links.

To link a file, like a PowerPoint presentation .ppt, or a Word document .doc, or an Excel spreadsheet .xls file, from the **Actions** menu, select **Link** ▶ and then **File....** Find the file in the directory window and click the **Insert**

A search of your Inbox will net all e-mail messages you sent to a client.

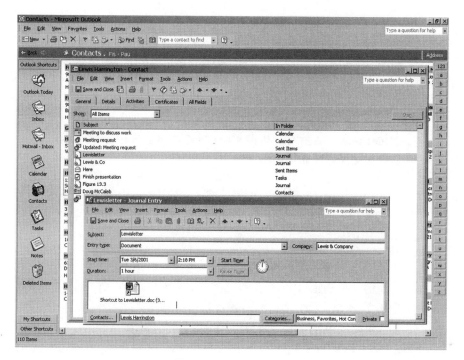

Figure 13.5

button. A Shortcut to the file is added to the contact's Activities list. Any file — not just Office XP documents — can be linked to an individual contact.

Contacts can also be linked together. At the bottom of the contact card is the button **Contacts...**. Just click it to select other contacts to be linked to that one. When you link one contact, the other is automatically linked back.

One other method of organizing is using the **Categories...** button. You can assign all Outlook items a category, so, let's say you want to keep track of all people, meetings, and e-mail messages related to Project Alpha, you can create a category to keep track of it.

Any file can be linked to an individual contact.

From the Outlook Folder List, you can also check in on the **Journal**, which helps track all of these items. It can also measure how long you spend on certain projects or tasks.

Finally, one little tip for Outlook. It's happened to us often. A colleague walks up and asks, "Hey, do you have Bob's phone number?"

As we confirm that Bob's contact information indeed resides in Outlook, we're asked, "Can you print it out for me?"

Rather than printing it, we send all the information along — Bob's entire contact card. From within Outlook 2002, in the **Contacts**, select the contact card (or cards by holding down the **CTRL** key). From the **Actions** menu, select **Forward** (or press **CRTL+F**). An e-mail is created with the contacts attached. Other **Outlook** users can drag and drop the cards into their Contacts within Outlook 2002.

> *A new feature in Microsoft Office XP takes cutting and pasting between applications to a whole new level.*

Cut-and-Paste Again and Again

A new feature in Microsoft Office XP takes cutting and pasting between applications to a whole new level. Office XP supports storing up to 24 items — text, images, objects of almost any sort used in Office applications. In fact, this new, robust **Office Clipboard** also integrates with the Internet Explorer Web browsers and some other Windows components.

While in Word 2002, open a document and choose some text to move somewhere else. Three different methods allow for cutting (or copying) that text. Our informal survey with colleagues and friends found all three methods were used, no matter what level of sophistication:

♦ To cut, hit **CTRL+X**. To copy, hit **CTRL+C**.

♦ Right-click and from the drop-down menu, select **Cut** or **Copy** as appropriate.

♦ From the **Edit** menu, select **Cut** or **Copy** as appropriate.

The first time this happens, you might notice a little window appears with a Clipboard icon that says, "1 of 24 – Clipboard. Item collected." The Clipboard icon appears in the Windows System Tray, similar to *Figure 13.6*.

Wherever in Windows you're working, you can copy items to the Clipboard.

Figure 13.6

Wherever in Windows you're working, you can copy items to the Clipboard. Windows itself stores a single item. You can paste this item into almost any application that supports it.

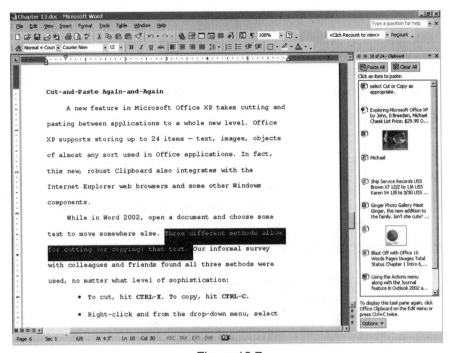

Figure 13.7

If you decide to paste the most recent item collected in the Office Clipboard, it's easy as before.

With Office XP, the Clipboard now supports two dozen items. When you highlight something else and cut or copy, the Clipboard Task Pane opens, as seen in *Figure 13.7*. You can also see from that image a wide variety of items on the Office Clipboard — an image from FrontPage, a chart from PowerPoint, text from a Web site and from Word, several cells from Excel.

If you decide to paste the most recent item collected in the Office Clipboard, it's easy as before. Find the place you want it and perform one of these three methods:

◆ Press **CTRL+V**.

◆ Right-click and select **Paste**.

◆ From the **Edit** menu, select **Paste**.

If you want to choose an item other than the one most recent, you can click on it from the Clipboard Task Pane and it will be pasted wherever your cursor is located.

You can also collect several items and paste them all at once by choosing the **Paste All** button at the top of the Clipboard Task Pane.

You can also delete an individual item from the Office Clipboard. Hold your cursor over the item and a drop-down menu arrow ▼ appears to the right. Click it and select **Delete**. Or clear the entire Office Clipboard by clicking the **Clear All** button at the top of the Task Pane.

Does the Office Clipboard Task Pane get in the way? Select the **Options** button at the bottom of the Task Pane when it appears and select **Collect Without Showing Office Clipboard**.

Working together can refer to how a team of people works in Office.

What is Team Productivity?

But working together does not just mean how Word works with Outlook. Working together can refer to how a team of people works in Office. Microsoft Office XP offers a lot in this arena, too. In Chapter 12, we examined SharePoint Team Services, which offers a platform to share information among a team.

Whether you're working with a large group of people or just one other, Office XP offers more tools to make the collaboration process easier.

The most often used is probably the review process in Word, where someone creates a document and sends it to someone else for editing and comments. Office XP makes this process a little more robust, providing greater control. Moreover, a single document can be sent to several people and returned back to the originator, who can reassemble all comments and changes into a single document.

And for those of you who have used the review process before, you may think it impossible to handle so many changes at once. But Office XP makes it possible — almost effortless!

An alternative to just sending in an e-mail message is a routing process.

The review process

Sending a copy to one person or a group is relatively easy. If you've finished your document and it's ready for review, from the **File** menu, select **Send To** ▶ and then choose **Mail Recipient (For Review)....** Word connects to Outlook and creates an e-mail message like *Figure 13.8*.

Address the message to everyone to route the document and send away.

An alternative to just sending in an e-mail message is a routing process. Routing provides a bit more comprehensive support. From the **File** menu, select **Send To** ▶ and then choose **Routing Recipient....** A window similar to *Figure 13.9* will appear. Select the appropriate people to whom the document should be sent. You can choose whether to route it to all reviewers or one at a time, one after the other.

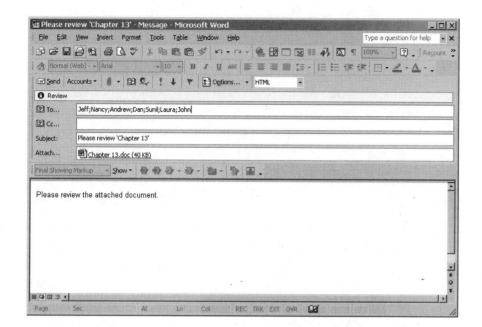

Figure 13.8

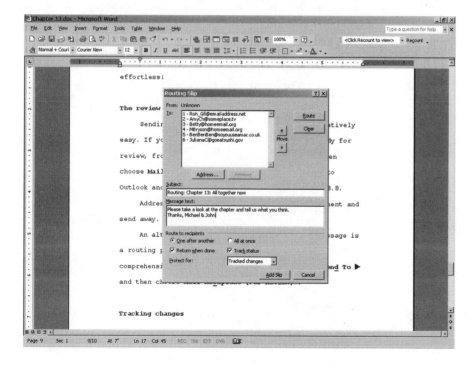

Figure 13.9

Tracking changes

Once a document arrives to a user, usually **Track Changes** is automatically turned on and the **Reviewing** toolbar is up. Word 2002 looks a little different than previous versions since deletions and changes appear along the right side of the document in colored boxes along with colored text, as seen in *Figure 13.10*. This fixes a common problem when more than a few people have reviewed a document and several make changes.

In previous versions, deleted text appeared with a strikethrough line.

In previous versions, deleted text appeared with a strikethrough line. It turned out hard to read. Now removing the text and putting it off to the side makes seeing the changes much easier. The "integrity" of the document remains — you can read it from beginning to end without skipping over deleted text.

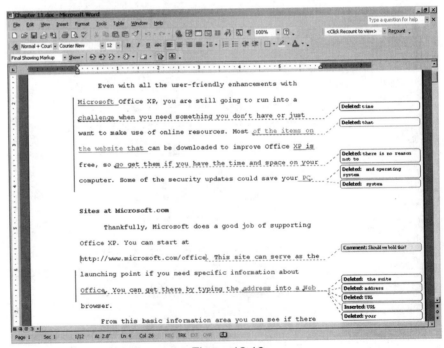

Figure 13.10

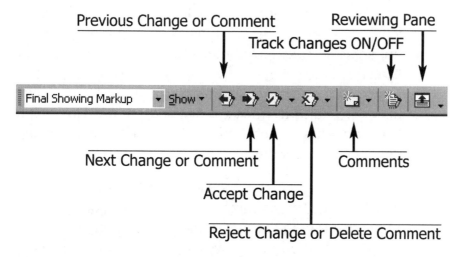

Previous Change or Comment   Reviewing Pane

Track Changes ON/OFF

Next Change or Comment   Comments

Accept Change

Reject Change or Delete Comment

Figure 13.11

When reviewing a document, add and make changes as you see fit. While Track Changes is on, Word takes care of everything, keeping up with who you are and what you're doing.

To turn Track Changes on or off, access the **Reviewing** toolbar (*Figure 13.11*) (right-click in the toolbar area and select **Reviewing** from the drop-down list). Or you can go to the **Tools** menu and select **Track Changes**. Or press **CTRL+SHIFT+E** to toggle it on and off.

While Tracking Changes is on, Word takes care of everything.

Putting notes into documents

Sometimes when reviewing a document, changing the text isn't quite right. A comment from "Great writing here!" to "What were you thinking?" to "I don't think these numbers are right" may be needed.

Comments to the rescue!

The Comment button on the Reviewing toolbar allows reviewers to insert suggestions to the author without actually making changes. Click the button that looks like one of the yellow sticky notes (*Figure 13.11*). If Track Changes is on, then the note will appear in a colored box off to the right. If Track Changes is off, the Reviewing Pane appears at the bottom of the window and the note can be inserted there.

Once you finish reviewing a document, just close it. Word 2002 recognizes that a document needs to move to either the author or the next person in the routing slip, as seen in *Figure 13.12*. You can also select the **File** menu, choose **Send To ▶** and then **Next Routing Recipient....**

The Comment button on the Reviewing toolbar allows reviewers to insert suggestions to the author without actually making changes.

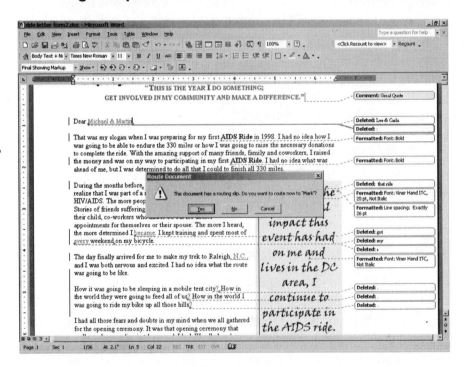

Figure 13.12

When routing, you may get a notice that the application is attempting to access e-mail without your consent. This is a security precaution built into Office XP. We've warned you about people writing viruses. This warning would let you know if a virus was sending out e-mail. But in this case, we're routing, so it's okay to click **Yes**.

Accepting/rejecting

Once a document has been routed to all appropriate reviewers, it returns to the author. The author can see, based on the colors, who suggested what comments. But if the colors get too numerous to track, just hover over each comment. The reviewer's name as well as the date and time of the suggestion or edit appear.

The author can go through a document and accept or reject each edit.

The author can go through a document and accept or reject each edit. Using the buttons in the Reviewing toolbar, click the icon with the blue checkmark to accept a change or the red X to reject it (*Figure 13.11*).

It's really as simple as that. Word 2002 keeps track of it.

Since you've got the hang of Word, now try it with Excel or PowerPoint. It's almost the same.

Wrapping Microsoft Office XP All Up

Writing this book has been quite a journey. Seeing the final version of Office XP is a bit like seeing a child finally grow up. When we started this guide, we had just installed the very first beta of Office. Even then, through all the quirkiness and crashes, we could see there was something about this suite that was special.

As we began the guide, new betas began arriving, with each one fixing some of the errors of the previous versions. We got the final code just as we were finishing up the book, just long enough to check to ensure that our experience writing about the many betas was still accurate.

As observers through every development step in this product, we are uniquely qualified to affirm that Office XP now stands supreme as Microsoft's best. It's a new beginning for Office suites, and the standard bearer that other products will be judged against.

We hope this guide has made your journey a little less stressful. Think of this book as a roadmap pointing out the important and "be-sure-not-to-miss" landmarks of the suite. But don't be afraid to travel down some side roads on your own. Office XP is nothing if it isn't friendly. Experiments with the applications will likely lead to discovery of new and exciting features and tricks. With so many customization options, our prediction is that this will be the first office suite to truly become personalized to each individual user.

And the true proof that Office XP is the right tool for your home or office? Well, it won over two jaded technical reviewers who, until now, had yet to agree on almost anything — except to disagree. We are confident that Office XP will win you over as well.

Experiments with the applications will likely lead to discovery of new and exciting features and tricks.

Index

Task Pane
   79, 83, 158, 177,
   198, 308, 309,
   331, 340
Task Panes 22
Tasks 323
template 43, 57
text icon 66
Text Wrap 175
Text Wrapping 71
Title Master 160
Track Changes 344, 346
Trojan horse 142
typewriter 7

U

username 263

V

vCard 141
Venn Diagram 183
virtual pen 66
virus 142, 302
virus, macro 303, 327
virus, self-propagating
   303
Visio 265, 269
Voice Command 294

W

W3C 242
Web query 316
Web Site Templates 230
Windows 2000 28
Windows 98 28
Windows Me 28
Windows Meta File 174
Windows NT 4.0 28
Windows System Tray
   339
wizard 50
WMF 174
Word Count 51, 53
word processor 7, 38
WordArt 73
WordPerfect 6
WordPerfect Graphics
   180
World Wide Web
   Consortium 242
worm 142, 303
WYSIWYG 9

X

XML 223, 320

Y

yellow 269

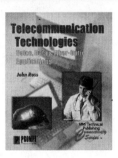

Computer Networking
for Small Businesses
by John Ross

Small businesses, home offices, and satellite offices have flourished in recent years. These small and unique networks of two or more PCs can be a challenge for any technician. Small network systems are vastly different from their large-office counterparts. Connecting to multiple and off-site offices provides a unique set of challenges that are addressed in this book. Topics include installation, troubleshooting and repair, and common network applications relevant to the small-office environment.

Exploring LANs
for the Small Business
and Home Office
Louis Columbus

Part of Sams Connectivity Series, *Exploring LANs for the Small Business and Home Office* covers everything from the fundamentals of small business and home-based LANs to choosing appropriate cabling systems. Columbus puts his knowledge of computer systems to work, helping entrepreneurs set up a system to fit their needs.

- Includes small business and home-office Local Area Network examples.
- Covers cabling issues.
- Discusses options for specific situations.
- TCP/IP (Transmission Control Protocol/Internet Protocol) coverage.
- Coverage of protocols and layering.

Communication
368 pages • paperback • 7-3/8" x 9-1/4"
ISBN 0-7906-1221-6 • Sams: 61221
$39.95

Connectivity
320 pages • Paperback • 7-3/8" x 9-1/4"
ISBN 0-7906-1229-1 • Sams 61229
$39.95 US

To order today or locate your nearest Prompt® Publications distributor
at 1-800-428-7267 or www.samswebsite.com

Prices subject to change.

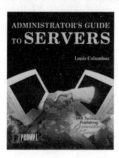

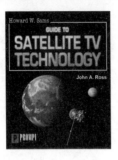